Build the Internet Speedometer. radioshack.com/DIT

PARTS

○ Getting Started with BeagleBone Black Maker Shed MSGSBBK2
○ RadioShack Tricolor LED Strip RadioShack 276-339
○ Enercell 12V/1500mA AC Adapter RadioShack 273-358
○ Enercell 2-Pin Adaptaplug™ Male to Tinned Leads (6")
  RadioShack 273-349
○ Solderless Breadboard Jumper Wire Kit RadioShack 276-173
  (Optional: jumper wires are included in the BeagleBone Black kit)
○ Breadboard RadioShack 276-003
  (Optional: Breadboard is included in the BeagleBone Black kit)

TOOLS

○ Computer for interfacing with the BeagleBone Black

②

Now that the wiring is complete and the BeagleBone has been booted up, login as root
and load the module uio_pruss:

`modprobe uio_pruss`

Change working directories to /lib/firmware:

`cd /lib/firmware`

Export the variables PINS and SLOTS with the following values:

`export PINS=/sys/kernel/debug/pinctrl/44e10800.pinmux/pins`
`export SLOTS=/sys/devices/bone_capemgr.9/slots`

NOTE: Rather than typing out the entire pathname at a command line, tapping the tab button autocompletes the path. Try typing in only the command below and after typing y, type tab:

`export PINS=/sy`

Autocomplete has its limitations, however, it can save a bit of typing.
Load the PRU by running: `echo BB-BONE-PRU > $SLOTS`
Verify the PRU is loaded by checking the following command's output.

`cat $SLOTS`

To test out the configuration, navigate back to ~/pypruss/examples/blinkled:

`cd ~/pypruss/examples/blinkled`

Run blinkled.py: `python blinkled.py`
The onboard LED should blink a few times, letting you know it is working.

③

Now that we have proper PRU access, we can setup
the speedometer code.
Change directories to ~/pypruss/examples: `~/pypruss/examples`
Clone the git repository: `git clone https://github.com/Make-Magazine/bb-led-internet-speedometer.git`
Change directory to bb-led-internet-speedometer:

`cd bb-led-internet-speed`

④

Run the Speedometer by running the python script pyGet.py:

`sudo python pyGet.py`

The pyGet script is configured to download a test file on speedtest.net's servers that is 10Mb. Under normal circumstances, a file this size will download quickly and the tricolor LED strip will be green. When the download is not as quick, the output color will be orange. And when the download speed is very slow, the output color will be red. Each time a 10Mb file is downloaded, pyGet.py notes how long it took and stores the result in speed.txt. Depending on the contents of speed.txt, one of three .bin files is invoked from the pyGet.py script: either red.bin, orange.bin, or green.bin.

NOTE: The speedometer code also contains a blue.bin and a other.bin. The blue.bin is simply an extra color that never gets used by pyGet.py, but is included. The other.bin is used as a failsafe in the conditional statements of pyGet.py.

NOTE: If you want to hack around the code — which is strongly encouraged — remember to run make in the bb-led-internet-speedometer directory or else the code changes will not be compiled into the .bin files

CONTENTS

COLUMNS

Reader Input 08

Welcome 10
Is Game of Drones the next X Games?

Made on Earth 12
Explore the amazing world of backyard technology.

FEATURES

Hands-On Health Care 16
When his wife was misdiagnosed, Michael Balzer used 3D printing and imaging to get her well.

Maker Pro Q&A 20
Meet the founder and president of TinyCircuits.

SPECIAL SECTION: DRONES

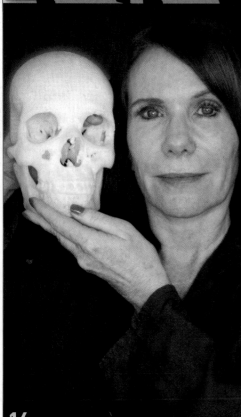

Rotary Club 22
Don't drone alone.

Formula FPV 24
Drone races are taking off.

The Multirotor Checklist 29
Remember all the little things that will keep you safely flying and having fun.

Drone Derby 30
Set up your own FPV race with these handy guidelines.

Hovership: 3D-Printed Racing Drone 32
There's no better way to start racing than to build your own FPV quadcopter.

Build Your First Tricopter 36
Tricopters fly smoother than quads and make better videos.

Noodling Around 42
This sturdy, low-cost airframe makes a great training quad. And it floats, too!

QuadH20 and WAVEcopter 43
A little moisture doesn't bother these waterproof drones.

Don't Be Dead in the Water 44
How to protect your electronics from the elements.

ON THE COVER:
Two Game of Drones "Hiro" quadcopters in FPV configuration race through the sky.
Photo: Hep Svadja

Make: Marketplace

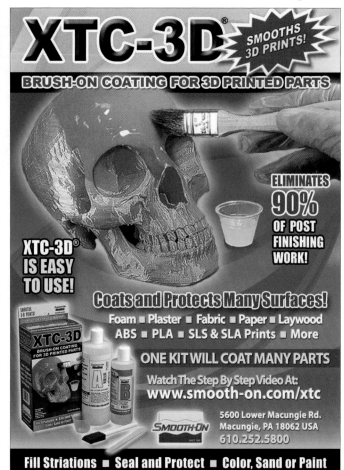

Make: Marketplace

PREPARE FOR BLAST-OFF

Make: Rockets
Learn how to build rockets from the ground up! Starting with the basics of rocket propulsion, readers will learn to make rockets from household objects and then move on to propellant-powered ones.

MAKERSHED.COM/MAKINGROCKETS

Making Makers
Is your child a Maker? How will you maintain their curiosity? This book shares stories and resources to inspire you and the kids in your life to create something amazing.

MAKERSHED.COM/MAKINGMAKERS

> NEW!
> MAKERSHED.COM

Maker Shed — The official store of Make:

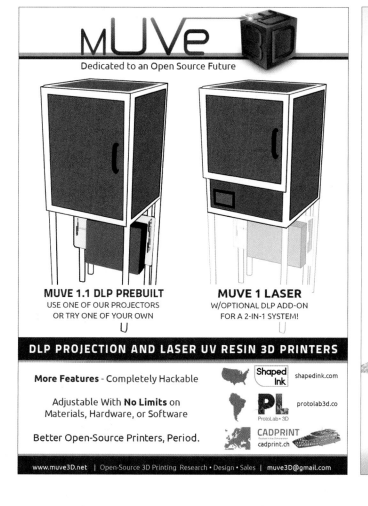

MUVe 3D — Dedicated to an Open Source Future

MUVE 1.1 DLP PREBUILT
USE ONE OF OUR PROJECTORS OR TRY ONE OF YOUR OWN

MUVE 1 LASER
W/OPTIONAL DLP ADD-ON FOR A 2-IN-1 SYSTEM!

DLP PROJECTION AND LASER UV RESIN 3D PRINTERS

More Features - Completely Hackable

Adjustable With **No Limits** on Materials, Hardware, or Software

Better Open-Source Printers, Period.

 Shaped Ink — shapedink.com
 ProtoLab • 3D — protolab3d.co
 CADPRINT — cadprint.ch

www.muve3D.net | Open-Source 3D Printing Research • Design • Sales | muve3D@gmail.com

---

PLASTIC INJECTION MOLDING MACHINES - STARTING AT $595!

Mold your own plastic parts. Perfect for inventors, schools or companies. Full details & videos on our webpage.

The affordable Model 20A turns your workshop drill press into an efficient plastic injection molding machine. Simple to operate and it includes a digital temperature controller.

No expensive tooling is required - use aluminum or epoxy molds.

The bench Model 150A features a larger shot capacity and is perfect for protoyping or short production runs. Capable of producing up to 180 parts per hour!

MADE IN THE USA

We also carry:
- MOLDS
- CLAMPS
- ACCESSORIES
- PLASTIC PELLETS

PayPal — VISA, MasterCard, [card], BANK

www.easyplasticmolding.com

IN 2013, PRINTRBOT RELEASED THE SIMPLE AND SHOOK UP THE CONSUMER 3D PRINTER MARKET. AT $299, it smashed the competitions' lowest price points and created a price war on Kickstarter. To achieve these cost savings, Printrbot created and adopted a few interesting tricks — some worked, some didn't. Surprisingly for such a budget machine, the Simple was still a good printer. Now Printrbot has redesigned the Simple, fixing some of those original problems and adding new features to improve this little workhorse even more.

REDESIGNED AND RENAMED

When Printrbot released the Metal Simple in late 2014, they renamed the wooden version as the Simple Maker's Kit. The new kit comes with a new price point too. At $349, the Simple Maker's Kit still pushes the bottom line. Although the entire machine has been redesigned, there are four major changes that really make the new version stand out from its predecessor: aluminum print bed, aluminum extruder, auto-level probe, and belt drives. These first two items were optional upgrades to previous models that mostly just improved the reliability of the printer. The aluminum bed didn't warp with changes in the environment like the previous plywood bed did. Printrbot's aluminum extruder is a simple extruder design that many are now mimicking due to its ease of assembly and reliability.

BELTS AND BETTER BED-LEVELING

The auto bed-level probe that Printrbot uses was first introduced in their Metal Simple line and is now being integrated into all of their machines. Unlike many bed-leveling probes that require contact with the surface to work, Printrbot has employed a novel system using an inductance sensor. The sensor can precisely measure the differences in magnetic fields caused by the presence of the metal print bed. This probe then measures multiple points on the bed surface — to determine how level it is to the print head, and how far from the nozzle the bed should be.

This system eliminates two of the largest problems most novice users have when getting started with 3D printing: bed leveling and nozzle height.

The original Simple used strings as its drive system, a method first introduced to 3D printers by the RepRap Tantillus. The implementation of the string drive on the Simple never quite worked right though, resulting in slipping or entirely unwrapped strings, potentially ruining prints half-way through. The Simple Maker's Kit now incorporates GT2 belts, a common timing belt used in many other printers. These belts are much easier to tighten correctly and keep the Simple running smoothly.

A FEW MINOR FLAWS

There are still some issues though. Although this is a great machine for beginners, most will find the 4×4×4" print volume quickly will become too small for them. There are upgrades to create a larger build volume, but they — at least to me — seem awkward and impractical. The machine seems to still suffer from some rigidity issues. There are times that even with the auto-level sensor doing its job, the front of the print can be squished down too tight due to drooping in the axis. Now that the bed is fixed and not adjustable, there is no way to troubleshoot this issue on the fly. Heat creep from the extruder can also result in poor print quality and extruder jams after long print runs.

CONCLUSION

All around, the Simple Maker's Kit is a great starter printer, especially for those who are still interested in building their printer and are on a tight budget to do so. This is a perfect machine for someone who wants a travel printer or one that is easy to take out to do demos. Although they no longer sell the wooden cases for these machines, Printrbot has uploaded the case designs to Youmagine so you can make one yourself.

For those who have older wooden Simples, don't fear. Printrbot offers a low-cost kit to upgrade your machine to the Simple Maker's Kit. The printer used for this review was my personal machine that was just upgraded with this package and it remains one of my favorites.

PRINT SCORE: 34

- Accuracy   1 2 3 4 **5**
- Backlash   1 2 3 4 **5**
- Bridging   1 **2** 3 4 5
- Overhangs   1 2 **3** 4 5
- Fine Features   1 2 3 **4** 5
- Surface Curved   1 2 3 **4** 5
- Surface General   1 2 **3** 4 5
- Tolerance   1 2 3 **4** 5
- XY Resonance   FAIL **PASS (2)**
- Z Resonance   FAIL **PASS (2)**

PRO TIPS

The Simple continues to be a leading machine for those who want to get involved in 3D printing on a small budget. The wooden Maker's Kit is one of the lowest-cost machines on the market while still creating great prints.

WHY TO BUY

After properly placing the auto-level tool, fine-tuning adjustments can be made in software and stored in the onboard memory.

How'd it print?

MATT STULTZ is a community organizer and founder of both 3D Printing Providence and HackPittsburgh. He's a professional software developer, which helps fuel his passion for being a maker. 3DPPVD.org

TOOLBOX | 3D PRINT REVIEWS

PRINTRBOT SIMPLE MAKER'S KIT

WRITTEN AND PHOTOGRAPHED BY MATT STULTZ

Solid updates further improve this budget performer

Printrbot Simple 2014 Maker's Kit
Printrbot.com

- **Price** $349
- **Build Volume** 100×100×100mm
- **Bed Style** Aluminum
- **Temperature Control?** Yes
- **Materials** PLA and other materials that don't require a heated bed
- **Print Untethered?** Yes, the board has a built-in microSD card slot
- **Onboard controls?** Not stock but an external screen can be added
- **Host Software** Repetier Host and Slic3r recommended, but any reprap host software can be used
- **Slicer** Slic3r, but any reprap slicing software can be used
- **OS** Windows, OSX, Linux
- **Open Software?** Third-party software
- **Open Hardware?** Yes

PRINT

GO, BRAIN: MAKING A SPLASH
By Carol Reiley
$30 : gobrain.com

Carol Reiley wraps a fun children's story around the concept of the growth mindset, and the result is a book that has lessons for both children and adults. Jason Pastrana illustrated the book beautifully, with bright and engaging artwork.

In the story, Lisa struggles at learning to swim while her brother Johnny is praised for his initial success. Johnny becomes too focused on this stage of learning to progress further. In contrast, Lisa is initially frustrated, but with the encouragement of her swim coach she works hard and achieves her goal.

The suggested age range is 3 to 7, but my 10-year-old liked it, and enjoyed reading it to his 5-year-old cousin. An illustrated parent's guide provides information and resources for developing a growth mindset. When kids are encouraged to get out of their comfort zone, their brains become stronger. Go, Brain!

— Andrew Terranova

NEW FROM MAKER MEDIA

MAKE: JAVASCRIPT ROBOTS
By Rick Waldron and Backstop Media
$29.99 : makershed.com

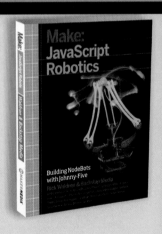

JavaScript Robotics is on the rise. Rick Waldron, the lead author of this book and creator of the Johnny-Five platform, is at the forefront of this movement. Johnny-Five is an open-source JavaScript Arduino programming framework for robotics. This book brings together 15 innovative programmers, each creating a unique Johnny-Five robot step-by-step, and offering tips and tricks along the way. Experience with JavaScript is a prerequisite.

NEW MAKER TECH

SPARK PHOTON
$19 : spark.io

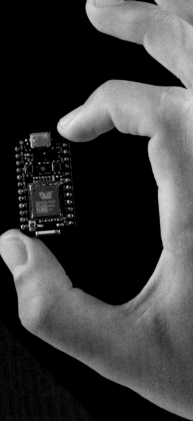

Spark enhanced their product line with Photon, their new Wi-Fi enabled hardware development board. The Photon improves on the popular Spark Core microcontroller by adding 802.11n Wi-Fi connectivity, more memory, and a faster ARM Cortex M3 processor. With its new SoftAP provisioning, setting up the Photon to connect to your Wi-Fi network is easier than ever and can be done within any web browser. Like the Core, the Photon fits in a standard breadboard for easy prototyping. Spark expects to start shipping the Photon in March 2015.

— Matt Richardson

RASPBERRY PI 2
$40 : bit.ly/raspberrypi-2

The Raspberry Pi 2, model B, is fast and inexpensive. Featuring a new Broadcom quad-core ARMv7 processor running at 900mhz and doubling the amount of RAM from previous Pi boards to 1GB, this is a must-have board for Pi enthusiasts and makers in general. The price remains the same as the Raspberry Pi B+, but this new Pi is at least six times faster, according to Eben Upton, CEO of the Raspberry Pi Foundation. During testing at the Maker Media Lab, we found that the new Pi could even function reasonably well as a desktop computer: We were able to slice STL files, browse the internet, and use GIMP — a popular open-source image editing tool — concurrently.

—David Scheltema

TOOLBOX

DREMEL 3000 SERIES ROTARY TOOL
$60–$80 depending on kit configuration : dremel.com

I have been using Dremel rotary tools for around 20 years for model making and general fabrication, and have always found them to be useful and dependable. The Dremel 3000 is no exception — I recently used it to remove support material from some 3D prints, and it performed as well as I've come to expect from Dremel tools. It even has some of the same design features as my first Dremel, including a positive locking collet assembly, a handy thumb button for operating it, and a variable speed switch to dial it in just right. Updates include a pleasing ergonomic shape, a collet wrench that is attached to the tool, and a router base-like cutter attachment with depth gauge.

The Dremel 3000 also comes with an assortment of accessories, including grinding stones, drum sanders, polishing tips and compound, and a few others in a handy case. Just for fun, I used the included cut-off wheel to cut some steel rod, and it made quick work of it. The manual offers useful information, such as maintenance tips and recommended RPM settings for various jobs.

I still have my first Dremel, and it works fine, although I have replaced the bushes a few times. This is easy to do with two easy-to-open brush caps on either side of the tool (another smart design feature that has been retained). I would recommend the Dremel 3000 to makers, who'll find it useful for model making, fabrication, prototyping, and electronics tasks.

— *Marty Marfin*

3DCONNEXION SPACEMOUSE WIRELESS
$130 : 3dconnexion.com

The SpaceMouse Wireless from 3Dconnexion is an exceptional tool for anyone who regularly works with 3D files. While it may not be worth the price for new CAD users — for those willing to shell out the cash, it is influential in speeding up the overall design process. The tactile buttons on each side of the mouse can be mapped to common commands in the program of your choice (of which the software supports many).

As its name suggests, the mouse is wireless, and the rechargeable battery lasts longer than I've had mine (weeks). My only gripe is that there aren't more buttons, as that would allow me to spend even more time away from the keyboard. For that, you'll have to spend more on one of 3Dconnexion's Pro models.

I can imagine myself using this mouse for years to come, and recommend it — or any of its more expensive siblings — to any 3D designer looking to take their speed to the next level.

— *Eric Weinhoffer*

BOSCH GLM 15 LASER MEASURE
$50 : boschtools.com

When we moved to our new *Make:* office, we needed to figure out the configuration of all of the editorial desks. Wanting to be helpful, I said, "No worries, I'll bring my tape measure in, and we'll figure out how much space everyone gets." I was practically laughed out of the office when I showed up with my 6' tape.

Laugh no more, as I recently discovered the Bosch GLM 15 laser measure. With one hand I can measure up to 50', at up to $1/8$" accuracy. I have a bunch of projects that require me to measure wall space, and with its design, I don't need another person to hold the other end (perfect for the single maker!). Other benefits are the continuous measurement mode, which allows you to walk off measurements from the wall and other surfaces, and its compact, square design that allows you to lay it on a flat surface to get accurate distances.

Throw it in your purse (or pocket), as I do, because you never know when you may need to measure how tall the street lights are (28' 6") or how much room the lame driver in the parking space next to you left you to squeeze into your car when you leave that nasty note. (13", jerk.) — *Cindy Lum*

[Bonus Tip: The GLM 15 is also great for sizing up rooms when house-hunting. It fits in your pocket and can be used a lot more discreetly than a tape measure. — SD]

FACOM CANTILEVER TOOLBOX
$115 (large) : ultimategarage.com

When filling up a large toolbox, you probably dump all your tools straight inside, right? This can make it a hassle to find and retrieve smaller hand tools from the rest of the jumble.

The solution: A multilevel, cantilever-style toolbox. I own a couple of cantilever toolboxes, and found Facom's to be the best. It's sturdy, well-made, opens easily, and provides a couple of options for internal organization. Facom's metal cantilever toolboxes come in several sizes, but the largest, 22"-long version is probably the most useful. Each has two foldout levels per side, and fold-down handles that help with opening and closing. You get the best of both worlds — easier organization of smaller items, and a larger compartment for bulkier tools.

— *Stuart Deutsch*

STANLEY STUBBY RATCHETING MULTIBIT SCREWDRIVER
$6 : stanleytools.com

If you're looking for a small but capable ratcheting screwdriver kit to start out with, you could do worse than the Stubby Ratcheting Multibit Screwdriver by Stanley.

For less than $10, you get a driver with a wide grip that will fit in a small toolbox. While a multitool might include screwdrivers, most hold the bit off-center, and will wear out your wrist quickly. This driver's ratcheting feature lets you twist it like a motorcycle grip, which is more comfortable and helps to drive in those long screws quicker.

The handle has room to store six bits, plus one in the business end. An assortment of Phillips and flatheads are included in the set, but I'd swap out some of them for torx bits.

In short, the Stubby Ratcheting Multibit Screwdriver is compact, capable, a great starter addition to a toolkit, and a great value to boot. Mine's been going strong for years.

— *Sam Freeman*

TOOLBOX

Gadgets and gear for makers
Tell us about your faves: *editor@makezine.com*

DJI Inspire 1
$2,900 : dji.com

The DJI Inspire 1 is quite a step up from the company's ubiquitous Phantom quadcopter — most notably in its physical design, with an ominous white fuselage mounted to an articulating carbon-fiber frame. The motor booms are angled down as landing gear during takeoff and landing, but automatically pull upward into a V-shape during flight to move them out of view and lower the rig's center of gravity.

The Inspire 1 sports large 13" props coupled to powerful brushless motors. Its pop-in-place battery carries 22 volts of juice — more than most cordless power tools. Downward-facing sensors optically track the terrain to help the craft stay in a fixed position during flight (especially useful indoors when GPS signals drop out) and help with soft, automatic landing and takeoff. And a bottom-mounted camera and gimbal rig allows for 4k video and 12 megapixel stills, capable of unobstructed views from any angle during flight. With its built-in Lightbridge wireless transmission capability, the footage is instantly viewable in high-definition on your radio-mounted tablet. It's not cheap, but its flight and video capabilities make this a must-have drone for any professional aerial videographer.
— *Mike Senese*

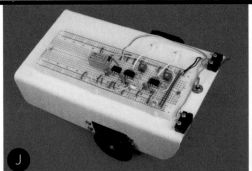

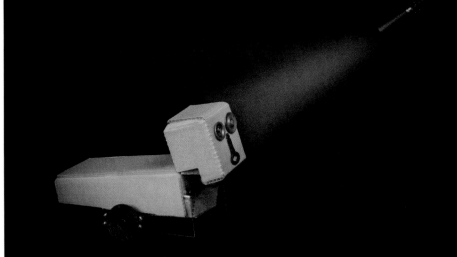

Figure I: Each servo is bolted into a notch cut in the frame. The wheels were included with the servo recommended for this project.

Figure J: The bare-bones cart in all its glory. Snap-action switches are mounted at the front to detect obstacles.

Figure K: The battery pack is secured with double-sided adhesive and is placed to apply maximum weight over the wheels.

snap-action switch can be our cheap-and-simple collision sensor, triggering a third timer in one-shot mode. This closes a relay for a few seconds, bypassing the photoresistors and forcing the cart to reverse away from the obstacle.

The complete schematic is shown in Figure G, and the breadboarded layout is in Figure H.

FABRICATION
Because this is a bare-bones cart, I didn't bother to make it look nice. The initial fabrication from a piece of ABS plastic is shown in Figure I, and the final cart is in Figures J and K. Note that the motors must both face the same way, because they'll be turning in the same direction.

Double-sided adhesive holds the battery pack under the frame. A DPDT switch controls the power, switching the 4.5V and 6V circuits separately.

Try to balance the cart with most of the weight over the wheels, so that the tail end slides easily across the floor.

TESTING
Use the cart in a dimly lit room. If it runs backward, you have too much ambient light. If it doesn't run straight, adjust the trimmer potentiometers. Now shine a flashlight at it, and it should stop and back up. If you angle the photocells in different directions, you can illuminate one of them but not the other, to make the cart turn. This works best on a smooth, hard floor that will allow the tail to slide to and fro.

UPGRADES
To increase the speed of the cart, make your own, larger wheels. To make it turn automatically when it backs up, add a tail wheel that pivots through a limited range, as I suggested for a simpler cart project in my book *Make: Electronics*.

To create some mystery, you could use infrared phototransistors with matching infrared LEDs in a handheld remote. Puzzle your friends by issuing verbal commands to the cart, while you control it secretly with your transmitter. ◯

+SKILL BUILDER   Your Obedient Servo

FINDING PHOTOCELLS

Cadmium sulfide photoresistors, also known as photocells, are not as common as they used to be. You want one that has a minimum resistance of 1K or less under bright light. The Silonex NSL4140 from Jameco Electronics should do the job, but if you try a photoresistor that has a higher minimum resistance (4K is typical), you can omit the 4.7K series resistors from the schematic.

The 25K trimmers allow you to fine-tune the photoresistor response. Note that photoresistors do not have a polarity.

Why not use phototransistors, which are more commonly available? Because they don't have such a smooth, gradual response to changes in lighting. You can try one in the basic circuit shown in Figure **F**, but I think it will tend to create an "all or nothing" response from your cart.

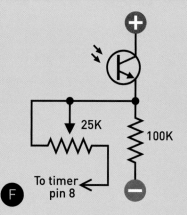

Figure H: A phototransistor can be substituted for each photoresistor, using a 100K resistor as shown here. Servo control may be rougher. Make sure the shorter lead connects with the positive power supply.

CART CONTROL

Suppose we have a cart with two wheels, each of which is powered by its own motor. This is a very common arrangement in bargain-basement robotics. When one wheel turns faster than the other, it steers the cart. When the wheels turn in opposite directions, the cart will pirouette.

If you remove the 100K trimmer in the test circuit and substitute a 25K trimmer in series with a photoresistor, you can steer the cart remotely with a flashlight. And if it bumps into something, a

WARNING
Be sure to include the 1N4001 protection diode across the relay, to suppress voltage spikes. Also, note that your relay may have different pinouts from mine. Check the relay datasheet to identify the pairs of contacts that close when power is applied.

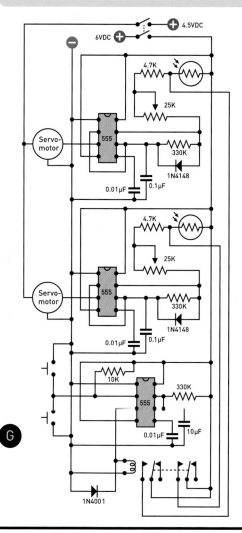

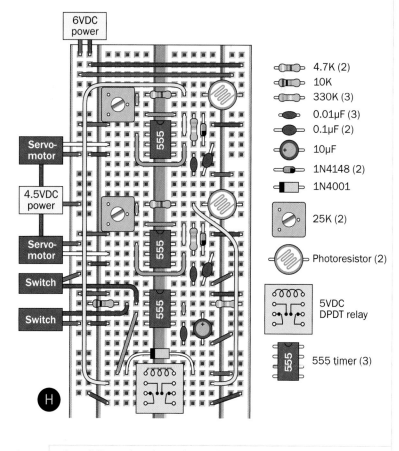

Figure G: The complete schematic for control electronics to run your obedient servo.
Figure H: Breadboard component layout for complete electronics.

86 makershed.com

makezine.com/44

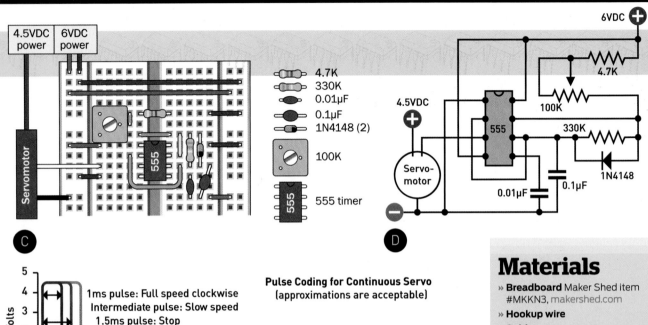

Figure C: When bench-testing a continuous servo using a 555 timer, the trimmer potentiometer controls its speed and direction of rotation.

Figure D: The bench-testing schematic.

Figure E: Three superimposed curves show the range of control pulses understood by a servo.

Materials

» **Breadboard** Maker Shed item #MKKN3, makershed.com
» **Hookup wire**
» **Solder**
» **Resistors:** 4.7kΩ (2), 10kΩ (1), 100kΩ (1), and 330kΩ (3)
» **555 timer IC chips,** original TTL type (3)
» **Trimmer potentiometers:** 25kΩ or 20kΩ (2), 100kΩ (1)
» **Capacitors:** 0.01μF (3), 0.1μF (2), and 10μF (1)
» **Diodes:** 1N4148 (2) and 1N4001 (1)
» **Relay, DPDT, 5VDC** any brand
» **Photoresistors, 1kΩ minimum resistance (2)** Silonex NSL4140 from Jameco Electronics, or similar
» **Switch, DPDT subminiature toggle**
» **Switches, snap-action (2)** also known as micro switch, with actuation lever, Honeywell ZM50E10E01 or similar
» **Battery holder, 4×AA**
» **Batteries, AA alkaline (4)**
» **Servomotors, continuous rotation (2)** Maker Shed item #MKPX18, SpringRC SM-S4303R, or any generic continuous analog servo
» **Wheels for servo, 1¾" diameter or larger (2)** if not included with motor
» **ABS plastic, ⅛"×12"×12",** Bend it with a heat gun. Or substitute ¼" plywood.
» **Bolts, #3×⅝" (12)** with nuts
» **Double-sided adhesive**

Tools

» **Multimeter**
» **Soldering Iron**

motor what to do.

Servos traditionally use 4.5VDC. The SM-S4303R is rated for up to 6VDC, but because you'll be running it for extended periods, and because some people may want to substitute a motor of their own, I'm going to stick with 4.5V.

SERVO CONTROL

Codes for controlling a servomotor can be generated by a plain old 555 timer. I'll use 6VDC to power the 555, because I want its output to activate a 5V relay. Obtaining the dual power supply of 4.5V and 6V, with a common ground, is easily done by tapping into a battery pack, as shown in Figure Ⓐ. You may need a 30W soldering iron to provide sufficient heat for attaching the wire shown in Figure Ⓑ.

The test circuit is shown in breadboard format in Figure Ⓒ, and as a schematic in Figure Ⓓ. When you turn the 100K trimmer potentiometer, the motor rotates forward or backward at different speeds. Great! But how does it work?

The timer generates a series of pulses, each lasting about 1 millisecond (ms) or longer, determined by the trimmer. A pause of about 18ms between each pair of pulses is established by the 330K resistor. For all continuous servos, a pulse length of 1ms means "turn clockwise," a pulse of 2ms means "turn counterclockwise," and 1.5ms means "stop." Intermediate pulse lengths make the motor turn more slowly. This is illustrated in Figure Ⓔ.

The high pulse, and the pause until the next one, should add up to a total cycle time that is always 20ms. But the 555 circuit doesn't work this way. When you lengthen the high pulse, the pause between pulses remains the same, so you actually increase the total cycle time. How can we correct this bad behavior?

Fortunately, we don't have to. It is a little-known fact that servos don't require accurate input. The motor won't care if a cycle is too short or too long, or if a pulse lasts for less than 1ms or more than 2ms. Therefore, a cheap 555 timer can control a servo just as successfully as a more-expensive microcontroller.

+SKILL BUILDER | Your Obedient Servo

EASY YOUR OBEDIENT SERVO

Written by Charles Platt

Learn the secret language of pulse modulation to master the continuous-rotation servo

CHARLES PLATT is the author of *Make: Electronics*, an introductory guide for all ages, and its sequel, *Make: More Electronics*. He is also the author of Volumes One and Two of the *Encyclopedia of Electronic Components*. Volume Three is in preparation. makershed.com/platt

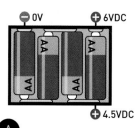

A
Four 1.5V batteries can supply 4.5VDC as well as 6VDC if you tap into the battery carrier.

B
The base of the fourth battery is usually available through an externally accessible rivet.

LITTLE ROBOT CARTS ARE WIDELY AVAILABLE IN KIT FORM, BUT BUILDING YOUR OWN CAN BE MUCH MORE REWARDING. You'll see for yourself how the components work, enabling you to customize your cart and even create a new version that's entirely your own.

This project is an ideal way to get started with simple robotics. It's so basic, it doesn't even need a microcontroller. Three timer chips provide enough intelligence to steer the cart in response to a flashlight and back up if it hits an obstacle.

SERVO HISTORY
Servomotors are the heart of this cart. Marketed originally for tasks such as moving the flaps on a model aircraft or the rudder on a model boat, they turn to a prescribed angle and then hold that position while waiting for further instructions.

You can control servos using off-the-shelf devices marketed by servo manufacturers, but many makers use microcontrollers for the job. The trouble with just downloading a block of Arduino code is that you won't understand what's really going on. The best way to learn is by bench-testing your own motor and driving it with your own hardware.

SERVO SETUP
For this project, I used SpringRC's SM-S4303R servo. It's relatively cheap and comes with a wide variety of parts to fit its shaft, including a wheel that will be ideal for our purposes.

You can use a different servo if you prefer, because they all understand the same codes. However, for this project it must be a *continuous rotation* servo, meaning that the shaft rotates continuously instead of just moving to a selected position and stopping there. (Catalogues don't always make this distinction clear, so shop with care.) Also, be sure to get a so-called "analog" servo, which is less expensive and less fussy about timing than the "digital" type.

All servos have three wires. The red and black wires supply power to the motor. A third wire of a different color carries the codes that tell the

the hook and start all of your measurements from the 1" mark. This works well and gives accurate results, as long as you remember to subtract 1" from your result. Trust me, no one who uses this method hasn't had a moment of dread after discovering something (or worse, multiple things) didn't fit to the tune of one extra inch. So stay awake out there. When choosing a tape measure, consider the type of work you are doing. If you primarily work with material shorter than 12 feet, don't buy a 25-foot tape. Those last 13 feet will never see daylight and the extra mass is heavy and cumbersome.

The folding rule (above) overcomes the hook problem by having a fixed metal cap at the end of its wooden rule. This makes for worry-free use, especially when measuring against something. It also has a nifty little sliding rule built into the end to measure depths and interior distances. On the downside, the thickness of the wooden blade means it must be laid on its edge to get accurate results and the way it folds creates a stair step shape that can make it awkward to use over distances.

The steel rule (at left) is a nice balance between the folder's consistency and the tape measure's small size, but its limitations are obvious. It's great for smaller work, but once you get beyond the 6" mark, one of the above will have to take over.

Honorable mention goes to the story pole or story stick. This is usually a long piece of wood that one puts their own marks on for transferring measurements. This can be more reliable because it gets rid of those numbers, and every distance is as marked. Story poles are very useful when you're measuring larger projects with multiple components (like a kitchen or library) or when you need to transfer the same dimension over many parts. It helps eliminate measuring mistakes.

SQUARES

For layout work, a square's primary function is to draw lines 90° perpendicular to a side. As always, there are a few types available but what sets them apart is what else they do. For me, a combination square (at right) is the most useful. Not only does it give me 90° and the occasional 45°, it also transfers measurements from one piece to another, finds the true center of a board, and checks depths and helps set up tools. It's hard to imagine woodworking without it. Definitely spend up when buying one. Get the best one you can afford. A loose, out of square or hard to move blade creates more frustration than it's worth.

The speed square is handy as well, but it is more suited to carpentry. I find the deeply stamped numbers to make for jaggy lines, so I use it mostly for rough layout and marking.

The *sashigane* is the standard square for Japanese joinery. It looks like a Western framing square but has a much thinner, flexible blade. And also like the framing square, it is covered in mysterious, oddly spaced numbers and strange markings that when in the right hands can be used to figure and lay out some pretty complicated joints. Since I have yet to decipher one, those hands are not mine.

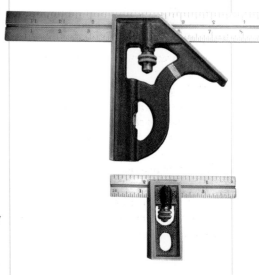

BUILD THESE SIMPLE WOODWORKING PROJECTS BY LEN CULLUM

WORKHORSES
Use a simple mortise-and-tenon joint to make these fine-looking shop horses that'll last a lifetime.
makezine.com/go/workhorses

SALT AND PEPPER WELL
So much more elegant than the usual cardboard shakers in your picnic basket.
makezine.com/go/salt-and-pepper-well

JAPANESE TOOLBOX
With its clever lift-out lid, this strong wood box carries my tools to nearly every big job.
makezine.com/go/japanese-toolbox

+SKILL BUILDER | Understanding Basic Woodworking Tools

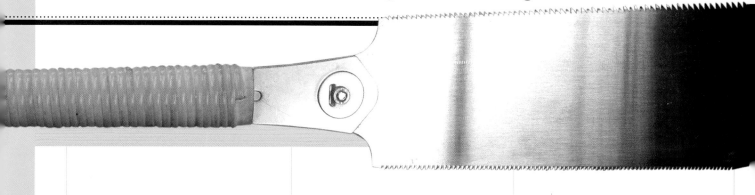

HANDSAW

As with the hand plane, much of the work a handsaw performs has been picked up by the powered version. Even so, the handsaw remains a useful and necessary part of a woodworker's collection. For cutting wood, there are two basic types: rip saws and crosscut saws.

Rip saws are meant to cut in the direction of the grain and typically have fewer, bigger teeth. Crosscut saws are, as the name implies, for cutting across the grain. They typically have more and finer teeth in order to shear the grain and leave a cleaner cut.

While general-purpose and combination saws exist, they tend to be a little too aggressive for careful work. My choice of handsaw is a Japanese *ryoba nokogiri* (double blade saw), shown above. It has rip teeth on one side, crosscut teeth on the other, and unlike Western saws, it cuts on the pull stroke. While they used to be difficult to find, you can now usually get them at home stores.

CLAMPS

Without clamps, nearly every operation with the preceding tools becomes more difficult. Not only are they good for holding together the final assembly, their ability to keep things where you want them while you work is invaluable. There is little that is more frustrating than trying to work a piece of wood that keeps sliding around. A couple of clamps, are essential and most woodworkers, at least once in their life, have repeated the mantra "you can never have enough clamps." Two 24" bar clamps, like the one shown below, are good. Four are better. Eight are better still ...

WOODWORKING PROJECT LAYOUT TOOLS

Accurate layout work is the critical first step to a successful project. Without precise, repeatable marks, it is very difficult to get everything to come together at the end. So now I'll go over some of the basic tools for measuring, marking, and transferring lines. My big three (actually four) tools for almost all of the work I do are the tape measure, a high-quality 12" combination square, a .005 drafting pen, and a 4" combination square for smaller work.

MEASURING LENGTH

The three most common measuring devices you're likely to find in a wood shop are the tape measure, folding rule, and steel rule. All three have their good and bad points. But as with all tools, find the one(s) that fit your style and make the most sense to you and the way you work.

The familiar tape measure with its spring-steel blade rolled up into a small box is fast and can measure distances that would require a massive folding rule. On the down side, the little hook at the end of the tape can introduce inaccuracy. When new, the hook slides on rivets just enough to adjust for the thickness of the hook's metal. When measuring to the inside of something, the hook is pressed in; when on the outside, the hook is pulled out, keeping the measurements accurate. This works great for a while, but over time, the holes and rivets can wear and get bigger, or the hook can be bent when the tape measure is dropped. To remedy this, most woodworkers "burn an inch." This is where you ignore

THE HAMMER

Nothing says blunt force like a hunk of metal on the end of a stick.

It's probably the oldest tool in the book. When I first started woodworking, I remember seeing a picture of a guy with his hammer collection, it was a whole room filled with hundreds of different hammers. At the time, I couldn't imagine needing more than one, but I feel much differently now.

Within eyeshot as I type this, I can see nine hammers. Each is different and each sees (fairly) regular use. The one pictured below is easily my favorite. It's a 375g Japanese carpenter's hammer. One face is flat, for driving nails, the other is slightly convex for driving the nail below the surface. I use it for everything from driving chisels and adjusting planes to knocking joints together and closing cans. It's my go-to hammer. The weight is right, and I like its balance.

If your work will require a lot of nailing, a claw hammer might be a better choice. Personally, I would just add a small pry bar to my collection.

CHISELS

Chisels can be used for anything from heavy chopping to light paring or fine carving. While also known to open paint cans, turn screws, and act as a pry bar, these are not recommended uses. Seriously, use a screwdriver. A screwdriver will appreciate the attention. While there are hundreds of chisel sizes and styles, most people can get by with four: ¼", ½", ¾", and 1" standard bench chisels.

There are virtually no chisels that are ready to use right off the shelf; they all need some sharpening to get them to sing. Once you experience a truly sharp chisel, you will understand the difference, not only by what you're able to achieve, but the ease with which you can do it. Below is a heavy patterned chisel called *atsu-nomi* (thick chisel) that's used for cutting joints in large timbers. It's part of a set made for me by master blacksmith Iyoroi, and it's one of my favorites.

HAND PLANES

Historically, hand planes were used mostly (but not exclusively) for smoothing and adjusting the thickness of rough board (called "thicknessing"). These days most stock dimensioning is done by machines, but this doesn't mean the hand plane is obsolete. It remains an incredibly useful tool that no woodworker should be without.

A well-tuned plane can do in minutes what can take a sander an hour, and produce an arguably better surface in the process — allowing you to work while standing in a pile of shavings instead of a cloud of dust. If I had to choose only one, it would be a low-angle block plane, pictured below. It can be used for everything from trimming and shaping stock to finish-planing surfaces. Like chisels, they're rarely ready to use out of the box and need to be sharpened before use.

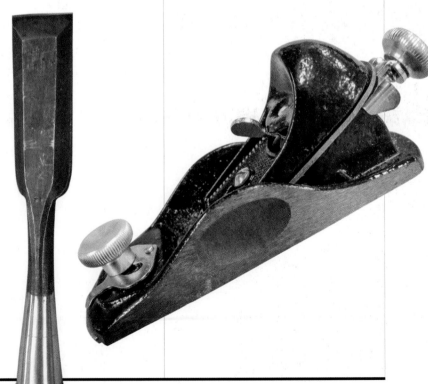

TIP
A tip for claw hammer users: When doing lighter chopping, try using the side of the hammer head to drive the chisel instead of the face. It gives you more control and a larger striking area.

SKILL BUILDER

Learning new tricks every issue
Tell us what you want to learn about: *editor@makezine.com*

EASY UNDERSTANDING BASIC WOODWORKING TOOLS

Ten simple hand tools for building almost anything   Written by Len Cullum

LEN CULLUM is a woodworker living in Seattle, where he specializes in building Japanese-style garden structures and architectural elements. When not woodworking, he teaches at Pratt Fine Arts Center, writes, and dreams of a robot that would sharpen his chisels.

I'm going to focus on what I consider the six basic hand tools for working with wood, plus my four go-to tools for measuring. These are the fundamentals that will allow you to build most anything. Keep in mind that no one tool is right for everyone. The hammer that I love might be the one that makes your wrist sore, or my favorite saw might feel backward. Don't be afraid to try different tools and techniques until you find the ones that feel right and make the most sense to you.

Can you turn your garage into a factory?
With a **ShopBot**, yes you can.

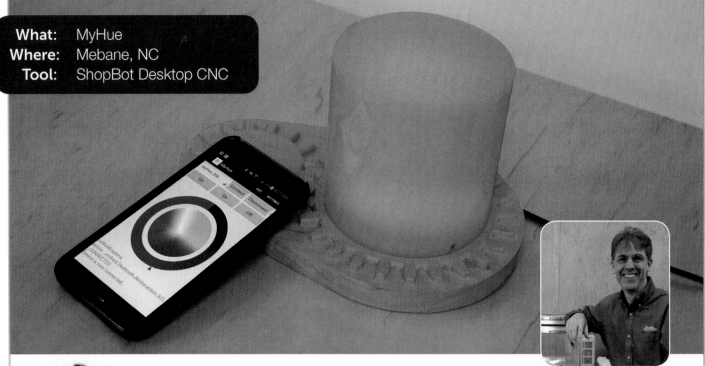

What: MyHue
Where: Mebane, NC
Tool: ShopBot Desktop CNC

Ryan Patterson is an entrepreneur. He's turned his home garage into a micro-factory for making his invention, the MyHue. It's a fascinating case study of the flexibility, power, and precision afforded by just one ShopBot Desktop, a small-footprint CNC tool.

Production Steps Using the ShopBot Desktop:

1. **Cut the acrylic lens.** The ShopBot is used to cut the lens profiles and to engrave different designs into them.
2. **Cut the wooden base.** First Ryan cuts a pocket for the electronics and cuts the overall profile of the base, then flips the piece over and cuts the pocket that holds the lens.
3. **Make copper circuit boards.** Ryan achieves this precision work in two passes: first he cuts the front side of one circuit board and the back of the other, then flips the pieces over and repeats the process.

Learn more about the MyHue and see photos at ShopBotBlog.com

"The idea for MyHue came to me because of my 'day job' at ShopBot. It's an open office space where you hear everyone's smartphone beeps, pings, and ringtones as texts, emails, and alerts come in. MyHue replaces this noise with light cues."

~ *Ryan Patterson*

Want to prototype or produce a new product? We offer the CNC tools and production support to help you configure the proper workflow. Give us a call!

We make the tools for making the future.

888-680-4466 • ShopBotTools.com

PROJECTS

Toy Inventor's Notebook
TINY TOY CARTESIAN DIVER
Invented and drawn by Bob Knetzger

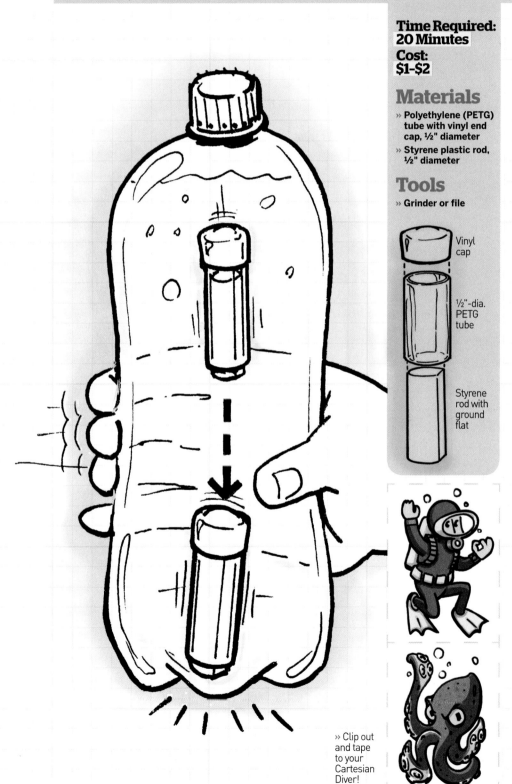

Time Required:
20 Minutes
Cost:
$1–$2

Materials
» **Polyethylene (PETG) tube with vinyl end cap, ½" diameter**
» **Styrene plastic rod, ½" diameter**

Tools
» **Grinder or file**

Vinyl cap

½"-dia. PETG tube

Styrene rod with ground flat

» Clip out and tape to your Cartesian Diver!

BACK WHEN CEREAL COMPANIES HAD BUDGETS to do such things, I got to invent some fun "free inside" premiums. It's a real challenge to come up with a toy idea that kids would like, moms would buy, advertising executives would get excited about, safety testers would approve, manufacturers could mold — and all for just a few cents each!

A premium I didn't invent but admired was "Diving Tony," a clever version of the Cartesian diver, a classic science toy. Inside a sealed bottle, a tiny Tony the Tiger mysteriously performed tricks, diving up and down. (Read more at makezine.com/go/cereal-science.)

You won't find a Diving Tony in your cereal bowl today, but you can make your own Cartesian diver toy quickly from just a few parts. I used a short piece of clear PETG tubing with a matching vinyl cap. I also cut a piece of solid styrene rod that fits snugly inside the tube. File or grind a small flat along the length of the rod. This provides a channel for the water to compress the bubble of air inside the cap. Cap the tube and insert the rod.

Test the diver in a sink of water for neutral buoyancy: Slide the rod in or out to adjust the size of the bubble air inside until it just barely floats to the top

Fill a 1-liter plastic bottle with water, insert your diver, and cap tightly. When you squeeze the bottle, the water pressure compresses the air bubble, which then displaces less water, and the diver sinks. When you release, the bubble expands, and the diver rises. With a little practice you can make your Cartesian diver obediently rise or dive on your command.

So why is this toy called a Cartesian diver? As Rene Descartes himself famously said, "I sink, therefore I am!" ◆

To see a demo video, visit the project page at makezine.com/go/tiny-toy-cartesian-diver.

number of times, but was not able to develop a repeatable process that would turn rubber into a useful raw material.

He kept trying until he had spent all his money, so he borrowed some. He spent that and borrowed more. Eventually, he and his family were broke and living on the charity of his friends. Things looked bleak.

THE LUCKY ACCIDENT

Then an accident changed things. On a cold winter day in 1839, Goodyear inadvertently brought a piece of rubber that he had treated with sulfur into contact with a hot stove. The inventor looked at the piece in amazement. The hitherto gummy yet fragile compound had become something wonderful.

Goodyear's rubber fragment did everything that natural latex could not. It had become, in modern parlance, *vulcanized*. It was tough and durable in hot weather, and stayed flexible in cold weather. This, Goodyear knew, was a product with a huge future. By the winter of 1841, things were looking up. Goodyear's new process was an astounding success, and money started to come his way.

But sadly Goodyear was not a capable businessman. While he did obtain a patent for the vulcanization process in 1844, he licensed the process at rates that were far too low for him to make money. Worse, when patent infringers stole his work, he spent more on his attorney's fees than he was able to recover from the pirates.

He spent the rest of his life attempting to make good his dream of becoming a millionaire rubber manufacturer. Goodyear staged magnificent displays showcasing rubber products including furniture, floor coverings, and jewelry at London and Paris exhibitions in the 1850s. But while in France, his French patent was cancelled and his royalties stopped, leaving him with outstanding bills he could not pay. Goodyear was thrown into a debtors' prison.

When he died in 1860, Charles Goodyear was $200,000 in debt. Posthumously, royalties on his process started to roll in. His son Charles Jr. later made a fortune manufacturing shoemaking machinery. It's a shame that Charles Sr. never enjoyed the financial success his invention ultimately provided for his family.

MAKE A RUBBER ERASER

You can re-create the process of vulcanization using a modern product to make your own rubber items, like these pencil erasers in a shape you might recognize. We think Mr. Spock would appreciate them.

1. Determine how much Pliatex, water, and vinegar are needed to make your erasers by calculating the volume of the mold. You need to use enough liquid ingredients to fill twice the mold volume for each eraser because the solidified rubber occupies considerably less space than the liquid ingredients.

2. Mix equal amounts of water and Pliatex in a bowl and stir until smooth (Figure A). You can add food coloring to the final product if desired.

3. Measure an amount of vinegar equal to the amount of water used and place it in a bowl.

4. Add the Pliatex-water mixture to the bowl with the vinegar (Figure B). Stir briefly until the solution congeals into a cheesy, soft mass.

5. Working quickly, place the rubbery mass into the eraser mold and press firmly (Figure C). Pour off the surface water and press firmly again. Pour off any pooling water that appears on the surface of the mold.

6. If you simply let the rubber harden in the mold, it will turn into a harder, tougher eraser.

7. If you want a softer, more flexible eraser, place the mold and eraser compound in a 300°F oven for 10–15 minutes, depending on the size of the mold (Figure D).

Remove and let it thoroughly dry (Figure E). Your eraser is ready for use!

From a chemical standpoint, what's happening is this: Pliatex is a partially vulcanized latex rubber compound called polyisoprene. It will naturally begin to coagulate and harden on its own unless it is stored in an alkaline environment. So, the manufacturer adds ammonia to keep the pH level high during storage. When you take the cap off the bottle, you'll smell the ammonia used to preserve it.

When you add vinegar to the solution, the acetic acid in the vinegar quickly lowers the pH, and the large polymer molecules in the latex come out of the solution to complete the vulcanization process and form a solid piece of rubber.

How do you depend on vulcanized rubber every day? Share your ideas on the project page:
makezine.com/go/vulcanized-rubber

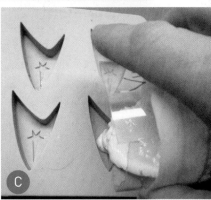

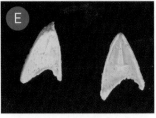

+ Rubber Band Powered Kits from the Maker Shed (makershed.com)

Bandit Rubber Band Sheriff Shotgun Kit

Cyclone Vacuum Cleaner Kit

PROJECTS

Remaking History

Charles Goodyear and the Vulcanization of Rubber

Written by William Gurstelle

Re-create the process that gave us car tires, rubber bands, and the boots on your feet

Time Required:
An Afternoon
Cost:
Under $10

Materials
- **Pliatex Mold Rubber** Pliatex is a compounded natural rubber latex product that's been partially vulcanized. It's available at craft stores and online. See sculpturehouse.com/s-278-mold-making-materials.aspx for information.
- White vinegar
- Water
- Food coloring (optional)

Tools
- Bowls (2)
- Measuring spoon
- Stirring spoon
- Candy mold

WILLIAM GURSTELLE is a contributing editor of *Make:* magazine. His latest book, *Defending Your Castle*, is available at all fine bookstores.

CHARLES GOODYEAR MAY HAVE BEEN THE MOST DOGGED AND UNRELENTING SOLO INVENTOR of the 19th century's golden age of invention. His was an interesting life, a wild roller coaster of ups and downs, although unfortunately mostly downs.

Goodyear was not wealthy or wildly successful. Many people are surprised to learn that he did not start, work at, or even know of the giant industrial concern called the Goodyear Tire and Rubber Company. The company, which was named in Goodyear's honor, was founded in 1898, about 40 years after Goodyear died.

THE EARLY YEARS OF RUBBER

In the winter of 1820, a rubber fad swept across America when millions of people bought rubber-coated boots to keep their feet dry. But the fad ended as abruptly as it started when consumers found that a single summer of hot weather turned their rubber shoes to mush. Natural rubber clothing just wasn't durable or practical.

At that point Charles Goodyear entered the scene. Goodyear thought that if he could figure out a way to toughen the rubber chemically he would have a product that people would buy. Although he knew almost nothing about chemistry, engineering, or business, he was, like the substance he was trying to make, resilient and tough.

He began to experiment with latex. He mixed in witch hazel, magnesia, even cream cheese in attempts to turn sticky, soft latex into durable, tough rubber. Nothing worked. He got close a

76 makershed.com

WIRELESS AND CLOUD COMMUNICATION

With the help of the wireless receiver and transmitter Bits, you can send signals without wires. And with the cloudBit, you can connect your project to a Wi-Fi network so that it can be accessed wirelessly via the internet. If you have a pair of cloudBits, you can have one pass its signal through the internet. This means your project can stretch across the globe.

WIRELESS TRANSMITTER AND RECEIVER

The wireless transmitter and receiver Bits work with each other in order to easily make your project work without need for the bits to be right next to each other. The transmitter communicates the incoming signals directly to the receiver, so it works without the need for a wireless network and without any kind of set up. These Bits are great because they're plug and play.

With the wireless Bits, you can make a remote control for a robot or car, create a wireless door lock, or be alerted in your living room when your back door is opened.

When you push the button on one circuit, the output should activate on the other. Try seeing how far you can go and still trigger the output with the button. The range of the wireless Bits is about 100 feet, so that it should still work from the next room, but your mileage may vary since there are a lot of factors that affect its performance.

CLOUDBIT

Not only does the cloudBit enable long distance communication between littleBits projects, but it can also connect to a multitude of online services through IFTTT (ifttt.com). And if you're an advanced developer, its open API means that the possibilities with how you can use it are limitless.

The cloudBit works by connecting to the internet over your Wi-Fi network. Through the internet, it establishes a connection to the littleBits Cloud Control servers, which receive signals from the cloudBit's input bitSnap connector and send signals to the output bitSnap connector. The littleBits Cloud Control server will help you connect your cloudBit to other cloudBits, to IFTTT, and to your own web servers by using its API.

GOING FURTHER: MAKE YOUR OWN BITS

There's a huge library of existing Bits available for you to choose from, but there are a few tools to help you create your own Bits. And through the BitLab program and the Hardware Developer Kit (HDK), you can even sell your creations. The HDK includes the proto modules, perf board, and bitSnaps to build your own circuit and add it to your library — or to the littleBits catalog itself.

Getting Started with littleBits is available this spring from Maker Media.

PROJECTS
Getting Started with littleBits

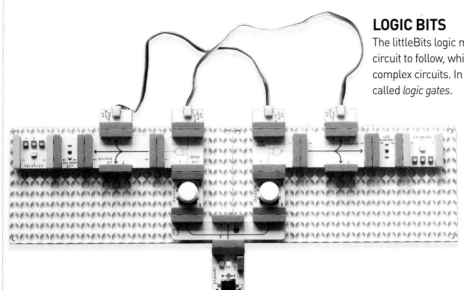

LOGIC BITS
The littleBits logic modules help you create rules for your circuit to follow, which opens up the possibility for more complex circuits. In electronics, these components are called *logic gates*.

What happens when you have multiple inputs and you want them to affect a single output? This is where the logic Bits are useful. The logic Bits work in the realm of digital, that is, only on or off. The important thing is that you can control two separate outputs, connect them both to a logic Bit, and connect an output to the logic Bit.

INVERTER
The inverter is one of the simplest wire modules, but it's very handy. There are cases when you want the output of a Bit to be the opposite: whenever it receives an ON signal, it outputs an OFF signal and vice versa. That's all the inverter Bit does. It outputs the opposite of its input.

DOUBLE AND
With a double AND Bit, both input signals must be ON in order for it to output an ON signal.

Input 1	Input 2	Double AND Output
OFF	OFF	OFF
ON	ON	ON
OFF	ON	OFF
ON	OFF	OFF

DOUBLE OR
With a double OR, when either input is ON, it will output ON. It's also important to note with the double OR that when both inputs are ON, it outputs an ON signal. The double OR is useful when you want two possible ways for a user to interact with your project.

NAND
NAND means "NOT AND." It does the exact opposite of the double AND. When both inputs are ON, it outputs OFF. In any other case, it outputs ON.

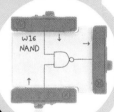

Input 1	Input 2	Double OR Output
OFF	OFF	OFF
ON	ON	ON
OFF	ON	ON
ON	OFF	ON

Input 1	Input 2	NAND Output
OFF	OFF	ON
ON	ON	OFF
OFF	ON	ON
ON	OFF	ON

NOR
As you may have guessed, the NOR Bit does the opposite of the Double OR Bit. In other words, when both of the inputs are OFF, then the NOR Bit outputs an ON signal. In any other case, it outputs OFF.

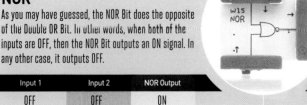

XOR
The XOR Bit means eXclusive OR. If either input is ON, the XOR Bit will output ON. But if both are ON, it will output OFF. If both inputs are OFF, it will output OFF. In short, it will only output ON when it receives a differing signal from its two inputs.

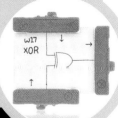

Input 1	Input 2	NOR Output
OFF	OFF	ON
ON	ON	OFF
OFF	ON	OFF
ON	OFF	OFF

Input 1	Input 2	XOR Output
OFF	OFF	OFF
ON	ON	OFF
OFF	ON	ON
ON	OFF	ON

WITH LITTLEBITS ANYONE, OF ANY AGE, CAN HARNESS THE POWER OF ELECTRONICS, MICROCONTROLLERS, AND THE CLOUD. You can combine these snap-together magnetic bricks to make simple electronic circuits or build robots and interactive projects that combine sensors and an Arduino-compatible microcontroller.

There's no better way to learn how to use littleBits than to jump right in and try them out. Here are the basics: how to power your Bits, how they connect — and a look at a few of the different inputs and outputs that you can use in your projects.

THE BITS

While there are over 60 different modules (or Bits) in the littleBits library to choose from, every module falls into one of four different categories, each with a particular color to make the modules easy to find and identify: Every Bit works with every other Bit in the library and it can keep growing to infinity.

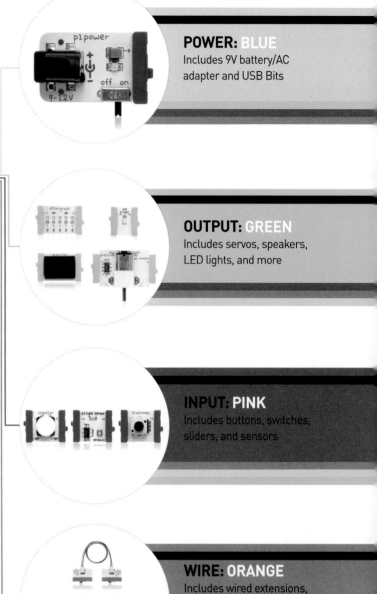

POWER: BLUE
Includes 9V battery/AC adapter and USB Bits

OUTPUT: GREEN
Includes servos, speakers, LED lights, and more

INPUT: PINK
Includes buttons, switches, sliders, and sensors

WIRE: ORANGE
Includes wired extensions, wireless CloudBit, splitters, inverters, and more

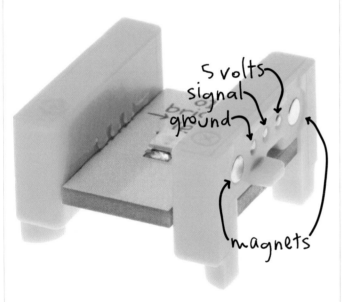

BITSNAP CONNECTORS

The Bits connect to each other magnetically with their bitSnap connector. This unique feature of the Bits helps you easily make the physical and electrical connections so that you can focus on creating your project. That means that there's no need to worry about soldering or making sure you're connecting the right wires.

If you look at the end of the connectors, you'll see that there are five metal pads. The two outer pads are actually the magnets that hold the Bits together. The three inside pads are electrical terminals. The middle terminal carries the signal, which is how the Bits communicate. The signal can be anywhere from 0 to 5 volts. 0 volts is an OFF signal, whereas 5 volts is an ON signal.

The signal terminal's voltage affects how the output Bits behave. For example, the more voltage there is on the signal line going into the LED Bit, the brighter the LED will be. The input Bits let you change the signal line's voltage and affect the Bits that come after it.

AYAH BDEIR is the founder and CEO of little-Bits. She's an engineer, interactive artist, co-founder of the Open Hardware Summit, a TED Senior Fellow and an alumna of the MIT Media Lab.

MATT RICHARDSON is a San Francisco-based creative technologist. He is co-author of *Getting Started with Raspberry Pi*, and the author of *Getting Started with BeagleBone* and *Getting Started with Intel Galileo* (available from Maker Media).

PROJECTS

Getting Started with littleBits

The new book from Maker Media will initiate you into the ecosystem of the versatile magnetic components

Written by Ayah Bdeir and Matt Richardson

Hep Svadja

PROJECTS

makezine.com/44

Make your own chocolate-hazelnut paste that's better than store-bought
DIY "Nutella" Spread

Written by Estérelle Payany

IMAGINE A RICH CHOCOLATE PASTE FLAVORED WITH hazelnuts, and one brand name comes to mind: Nutella. It's traditionally a store-bought indulgence — but making it yourself is a heck of a lot more fun than buying it off the shelf.

Here's how to make your own 100% natural version that's scarily good. Sure, it's almost a sinful explosion of calories, but once you've tasted it, it's impossible to resist.

CHOCOLATE-HAZELNUT SPREAD

1. Heat the oven to 350°F (180°C).

2. Spread the hazelnuts in a single layer in a shallow metal pan and roast for about 15 minutes. Remove the pan from the oven and let the nuts cool for 5 minutes.

3. When cool enough to handle, rub the hazelnuts between your fingertips to remove the skins. (You could also rub them with the folds of a clean dish towel.) Transfer the nuts to a food processor or blender and pulse for 2 to 3 minutes or until reduced to a fine powder.

4. In a saucepan, heat the chocolate and evaporated milk over medium heat until the chocolate melts. Stir occasionally to blend.

5. Pour the hot milk mixture over the ground nuts and add the cocoa powder, oil, vanilla, and salt. Pulse several times or until the paste is smooth and evenly colored. Spoon the paste into a sterilized 8oz jar. Seal and refrigerate. The spread will keep for up to 2 weeks.

➜ *This recipe is excerpted from* Better Made at Home: Salty, Sweet, Satisfying Snacks and Pantry Staples You Can Make Yourself *(Black Dog & Leventhal, 2014)*

Ingredients
MAKES: 8oz jar
- 4oz (10g) whole unsalted hazelnuts in their skins
- 4oz (100g) milk chocolate with hazelnuts broken into pieces
- ⅓ cup (9cl) evaporated milk
- 1 tsp unsweetened cocoa powder
- 1 tsp sunflower or nut oil
- ½ tsp pure vanilla extract
- Pinch of salt

NOTE: You can use sour cream thinned with a little milk in place of the evaporated milk.

Time Required: 30 Minutes
Cost: $5–$10

ESTÉRELLE PAYANY is a best-selling French cookbook author, culinary journalist, and food blogger.

Share your recipe tweaks and tips at makezine.com/go/diy-nutella

Guillaume Czerw

**Time Required:
2–3 Weeks**

**Cost:
Free**

What You'll Need:

» **Adult fig tree**, healthy and productive
» **Pruning shears** aka rose cutters
» **Permission** from the fig tree's owner to cut a branch or two
» **Glass vase or plastic bags**, transparent
» **Planter pots and potting soil**

this will help you decide if the tree will do well in your climate zone. Gardening books and websites (figs4fun.com is a great resource) will tell you where different varieties grow best.

The green-skinned King fig is a favorite for coastal, cool areas, while the Black Mission is suited for places with blazing summers. The King makes only an early crop — in June or July. The Black Mission makes two crops, a small one in early summer and a large crop in the fall. Other figs make only the later crop. The Calimyrna fig, which requires a unique insect to pollinate its fruits, may make no figs at all outside of California's Central Valley.

The best way to select the right variety for you is to locate a tree that's thriving — and bearing delicious fruit — in your neighborhood.

2. TAKE CUTTINGS

Cloning a tree requires wood — that is, branch cuttings. For best results you want healthy wood from the previous season — so at least 1 year old. The green wood near the branch tips may not root properly. This means you may need entire branches 36" long or more if you're taking cuttings late in the season. If you're cutting early in the spring, when the tree is just waking up after winter dormancy, you may only need 8" or so.

3. TRIM THE CUTTINGS

Snip off the leaves and any unripe fruits (Figure A). Cut each branch into 8" sections (Figure B). These are your future trees.

4. INCUBATE AND ROOT THEM

Figs are tough — adaptable and durable — so prompting your tree branch to send out roots and leaves will require wonderfully little finesse. I've had success just sticking a cutting into a pot of soil and, eventually, watching the twig become a tree.

Consistently better luck can be expected, however, by keeping the cuttings damp, or even submerged in water, until they sprout. Place the cuttings in a vase of clean water (Figure C) or seal them in clear plastic bags. If mold appears in a bag, open it and dry out the twigs for a few hours. You may, at this point, place them in water, where fungus is less likely to proliferate.

After 10 to 20 days you should see white, noodle-like root tips emerging from the wood (Figure D). Some cuttings will send out green leaves at the same time, or even before the roots appear. Whatever the sequence, once it's clear that a cutting is alive and raring to grow, give it some soil. Plant it 4" deep in a small pot (Figures E and F). Root hormones are not necessary.

5. GROW THE TREE IN A SUNNY PLACE

The more direct sun a tree receives, the sweeter its fruits will be — a simple equation of energy in, energy out. So, select your tree's planting location very carefully.

If in doubt, just transfer the small tree to increasingly larger pots, move as needed through the season. Eventually, you may find a good permanent location; in the meantime, you can expect a potted tree to produce luxuriant crops — though the tree's size will be restricted if it isn't planted in the ground.

REMEMBER: If you plant your tree in the front yard, you're likely to someday receive visitors kindly asking for a few inches of wood. Give it to them.

Share your garden-cloning creations at makezine.com/go/cloning-the-fig.

PROJECTS

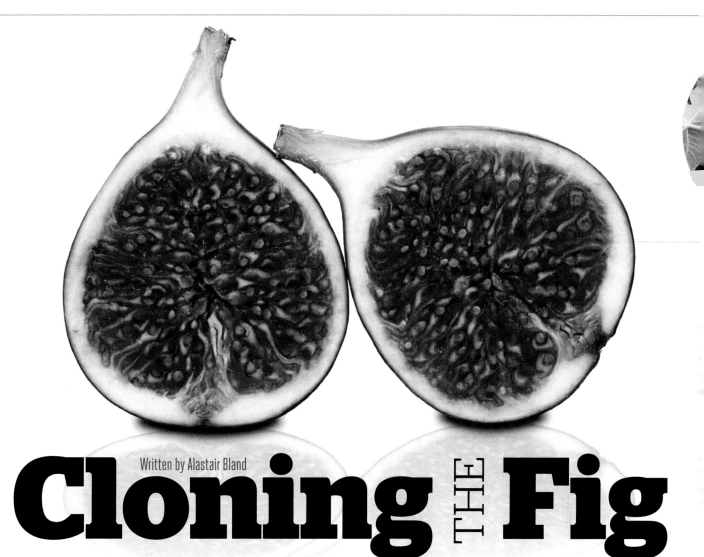

Cloning THE Fig

Written by Alastair Bland

Don't steal fruit from your neighbor's fig tree — clone it!

ALASTAIR BLAND is a freelance writer in San Francisco, where he was born. He reports on agriculture, science, fisheries, and the environment for publications including *Smithsonian*, *Yale Environment 360*, and NPR. He travels by bicycle, likes camping in bear country, ferments things, and loves figs.

THERE'S SOMETHING UNIQUELY BEAUTIFUL — BOTH VISUALLY AND CULTURALLY — ABOUT FIGS. The jammy, teardrop-shaped fruits have been loved for centuries in Europe, and for millennia further east. Now they're gaining popularity in the United States and Canada, where more and more people are growing the trees at home.

If you start watching over neighbor's fences, chances are, whether you're in Seattle, New York or Austin, that you'll start noticing fig trees. The temptation to duck into a yard and snatch a few low-hanging beauties can be powerful — but the civil option is to knock on the owner's door and, hopefully, gain permission to take away a bag full. Still, you'll eventually have to face the fact that these figs are not yours.

Unless you clone the tree. It's not rocket science. In fact, it's beautifully simple and one of the very oldest tricks in agriculture.

Happily, figs are among the easiest of fruit trees to clone. That's because they needn't be grafted onto an existing tree, as with most other fruits. Rather, a freshly cut fig branch can be crudely placed in the ground, where it will root, grow, and eventually produce heaps of the very same fruit borne by its mother tree. Fig trees can also be grown to fruit-bearing size in a pot — meaning you don't even need a yard.

1. SELECT YOUR FIG TREE

The tree should be a favorite that makes excellent fruit consistently. If its owner can tell you the fig variety — whether White Genoa, King, Black Mission, Kadota, or a hundred others —

NOTE: The cable's conductors face toward the "Camera" label on the board.

THE RASPBERRY PI COMPUTER IS AT THE HEART OF A LOT OF REWARDING PROJECTS — especially when you add the Pi Camera Module into the bill of materials. The module is easy to set up (see our Skill Builder tutorial at makezine.com/go/skill-builder-raspberry-pi-camera-module) and recently helped me with a home-security solution.

The Pi Model B+ is the perfect hardware for a security camera project: It's inexpensive, physically smaller and thinner than previous Pi models, and needs little power — the B+ can run for a very long time simply on power from a cellphone charger battery. And Calin Crisan's free software package, motionPie, takes all the hard work out of building a custom surveillance-camera system. The easy-to-use interface lets you check your cameras from anywhere, via a web browser on your computer, tablet, smartphone, or almost any internet-accessing platform.

CONNECT AND CONFIGURE YOUR PI

Download the latest stable motionPie image from github.com/ccrisan/motionPie/releases, unzip it, and write the image to a microSD card using your favorite image-writing utility. My current favorites are Win32DiskImager for PC and ApplePi-Baker for Mac.

Insert the freshly burned microSD card, with adapter, into the Pi B+, along with the camera module, Wi-Fi USB dongle, and Ethernet cable.

Finally, power up the Pi by plugging it in. The power LED will immediately illuminate, though on first boot the Pi will take a bit of time to initialize (Figures A and B).

FIND THE PI'S IP ADDRESS

Now you'll need to determine the IP address of the Pi. The easiest way I've found to do this is with an app called Fing for iOS or Android. Make sure your mobile device is connected to the same network as your Pi and run Fing to get a listing of all the IP addresses on your network. If you don't have an iOS or Android device you can always log into your router and look at the DHCP assignments.

Look through the Fing output for a line with "Raspberry Pi Foundation" and the IP address to its left; this is the address you'll enter into a web browser to access your Pi. You should be greeted with an interface and a live video feed.

SET UP YOUR CAMERA(S) WITH MOTIONPIE

Using the motionPie website, click on the key graphic in the upper left-hand side of the site. This will bring up a dialog box requesting a username and password. The username is `admin` and you can leave the password field blank.

Click the slider icon again to bring up the Admin menu, turn on Show Advanced Settings, change your admin password, and click Apply to save the settings. You'll then be asked to log back in using your updated credentials.

Finally, in the Admin menu, click the slider next to Wireless Network to the On position. Type in your network SSID name and password and click Apply to save. Setup is complete (Figure C).

GOING FURTHER

MotionPie includes additional configurations you'll definitely want to check out. The system can do motion detection and time-lapse videos. You can set a recording schedule so the system only records video during certain times of day, add more camera nodes — Pi NoIR night vision cams, USB cams, and networkable IP webcams — and even set up email alerts, which you can easily convert to text messages using your cellular carrier's email-to-text services.

Thanks to the small form factor of the Pi and Camera Module, this project fits almost anywhere. Consider deploying yours in a stuffed animal, a birdhouse, a live-streaming "street view" rig atop your car, or even concealed in a shirt — the possibilities are endless!

Time Required: 30–90 Minutes
Cost: $70–$100

Materials
» **Raspberry Pi Model B+ single-board computer** Maker Shed item #MKRPI5, makershed.com. You can swap out the Model B+ for the even smaller A+ (Maker Shed #MKRPI7) once everything is configured on the microSD card.
» **Raspberry Pi Camera Module** Maker Shed #MKRPI3
» **Raspberry Pi NoIR Camera Module** Maker Shed #MKRPI6
» **USB Wi-Fi adapter** Maker Shed #MKAD55
» **MicroSD card with SD adapter**
» **USB power supply**
» **USB cable, Standard-A to Micro**
» **Ethernet (optional)**

Tools
» **Computer with MotionPie software** free from github.com/ccrisan/motionPie

MICHAEL CASTOR is the Product Innovation Manager for the Maker Shed (makershed.com). His hobbies include skiing, martial arts, CNC machining, 3D printing, and flying drones.

REMEMBER: With great power comes great responsibility. Please use with good intentions only.

makezine.com/44

PROJECTS

1 2 3 How to Tattoo a Banana

Written by Jason Poel Smith

DON'T JUST PLAY WITH YOUR FOOD — MAKE ART WITH IT. I started tattooing bananas after seeing similar work by artist Phil Hansen. It is a great way to add some fun to packed lunches, and it doesn't ruin the food.

When you puncture or bruise a banana peel, the ruptured cells release chemicals that start to oxidize and turn brown. By using a fine-tipped needle, you can make detailed drawings on a banana peel. You can even use stencils to make a copy of your favorite picture.

You will need:
» Banana
» Image
» Scissors
» Scotch tape
» Needle
» Mechanical pencil (optional)

JASON POEL SMITH is a lifelong student of all forms of making, from electronics to crafts and everything in between. He creates the "DIY Hacks and How Tos" video series for *Make:* at youtube.com/make.

1. CREATE A STENCIL
Find a picture that you want to use and scale it to the appropriate banana size. Then print it out. Cut out the picture, leaving some blank paper around the edges. Next, use Scotch tape to attach the picture to your banana.

2. OUTLINE THE PATTERN
Using a needle, poke holes through the paper along all the major lines. Try to keep the holes close together and as shallow as possible. To make this easier, try attaching a needle to a mechanical pencil. When you're done, remove the stencil. You should see a dotted outline of your picture.

3. FILL IN THE DETAILS
Now you need to connect the dots and apply shading. Go back over all the major lines and make more dots to fill in the gaps. To apply shading, gently tap the surface to make very light scratches. The details will darken over time.

It's art that you can eat! ◉

See more banana masterpieces and share yours at makezine.com/banana-art

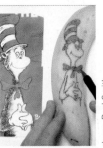

Jason Poel Smith

PROJECTS
Amateur Scientist

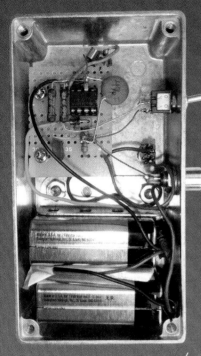

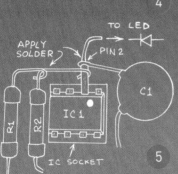

used both methods and much prefer the external method described here for initial experiments. This allows you to try various LEDs and collimator lengths. After you find the optimum combination, you can install the system inside an enclosure.

Two 9-volt batteries connected in series power the photometer. Because IC1 must not be powered by more than 16 volts, the 18 volts from the two batteries is reduced to 16 volts by Zener diode D1. This arrangement provides the maximum possible output voltage range for a data logging multimeter. If you plan to use a DIY or commercial data logger instead (see Part 2 of this project in our next issue, *Make:* Volume 45), the circuit's output voltage must not exceed the data logger's allowable input voltage. This is typically 5 volts, which means you can power the photometer with a 6-volt battery instead of D1, R3, and the two 9-volt batteries. Both power options are shown in Figure 3. If the logger's input must never exceed 5 volts, insert a 1N914 diode between the positive battery terminal and IC1.

ASSEMBLE THE CIRCUIT
The circuit is built on a 1½"×1¼" perforated board with copper traces on the bottom. The prototype board is shown in the open view of the photometer in Figure 4.

For best results, the op-amp's input (pin 2) should be isolated from the circuit board to prevent dust, fingerprints, and even the board itself from altering the gain of the op-amp. Isolating pin 2 in free air eliminates this problem. The easiest way to do this is to provide an 8-pin IC socket for IC1. Before inserting IC1 into the socket, bend pin 2 straight out so that it doesn't touch the socket when the other seven pins are inserted.

The next two steps are tricky, so refer to Figure 5 and take your time. First, solder the input side of R1 and C1 directly to pin 2. Then solder a wire directly between pin 2 and the LED cathode (–) terminal of the phone or audio jack.

INSTALL THE CIRCUIT IN AN ENCLOSURE
After the circuit board is assembled, clean the surfaces of R1, R2, and IC1 with a cotton swab dipped in alcohol. Then install the circuit board atop a pair of insulated standoffs inside a metal enclosure as shown in Figure 4. The photometer described here was installed in a Bud Industries CU-124 enclosure. A larger enclosure can be used, as can various metal containers sold by craft stores. If you use two 9-volt batteries (as shown in Figure 4), secure them in place with an angle bracket.

Figure 6 shows a pair of LED-collimator assemblies made from a gas coupler fitting and an aluminum or brass tube. The LED is inserted into an LED socket (optional) soldered to ⅛" phone plug terminals or soldered directly to the phone plug. [IMPORTANT: Be sure to observe polarity.] The phone plug is pushed into the open end of the lower coupler fitting and secured in place with a rubber O-ring.

You can omit the gas coupler if the plug's cap will slip over the LED, which means its leads must be clipped close and carefully soldered. Insert the open end of the plug into an appropriate collimator tube. A collimator length of 3"–4" should provide a field of view of around 5 degrees. As shown in Figure 6, the length of the collimator can be increased or decreased with a short length of heat-shrink tubing.

USING THE PHOTOMETER
On a clear day 10 minutes before sunset or 45 minutes before sunrise, place the photometer facing straight up on a level surface outdoors well away from light sources. A bubble level mounted atop the photometer will simplify alignment. If necessary, use shims to level the photometer. For best results, connect the photometer output to a data logging multimeter or a standalone DIY or commercial logger (Onset 16-bit HOBO UX120 or similar) and record data at 1-second intervals. If you don't have a logger, read the output voltage manually from a multimeter at 10- or 15-second intervals and enter the exact time and output voltage into a notebook or audio recorder. Automatic logging is preferable, but the manual method has been used for half a century.

GOING FURTHER
The raw twilight signal should provide a smooth curve when plotted on a graph of time vs. signal. Far more significant is a graph that plots the rate of change in the data against the elevation of the twilight glow. Part 2 of this project in *Make: Volume 45* will explain how to process your data so that it accounts for these parameters and reveals the altitude of aerosols over your location. Meanwhile, you can learn much more about the science of twilight photometry at makezine.com/go/twilight. ◆

Share your build and twilight observations at makezine.com/go/twilight-photometer.

Volcanic aerosols

HOW IT WORKS

The twilight glow straight overhead is very dim, and the photocurrent it generates in an LED is very small. Therefore it's important to use an LED installed in clear epoxy. For best results, use an LED that projects a narrow beam when used as a light source. (For details about using LEDs to detect light, see this column in *Make:* Volume 36.)

The photometer circuit is shown in Figure ③. In operation, the tiny LED photocurrent is amplified billions of times and transformed to a voltage by IC1, a TLC271BIP operational amplifier with a very high-resistance feedback resistor consisting of R1 and R2 in series. Capacitor C1 suppresses oscillation. The combined resistance of R1 and R2 controls the voltage gain of the amplifier. I have obtained best results using 40-gigohm resistors for both R1 and R2. When only 40 gigohms is required to provide a usable output signal during the 30–45 minute twilight period, switch S2 is closed to bypass R2. High-value resistors can be expensive and difficult to find, but I've had good results with Ohmite resistors from Mouser Electronics (mouser.com) and Digi-Key (digikey.com). For example, Mouser offers Ohmite's 40-gigohm axial-lead resistor (MOX-400224008K) for a reasonable $4.19 each. If 40-gigohm resistors aren't available, use 30- or 50-gigohms.

PLANNING THE PHOTOMETER

The twilight photometer should be installed in a metal housing to block electrical noise from power lines and radio signals. I learned this lesson the hard way while testing my first twilight photometers at Hawaii's Mauna Loa Observatory. The LED can be installed inside the enclosure with a small hole to admit the twilight glow, or inside an open-ended phone or audio plug fitted with a collimator tube and inserted into a jack atop the enclosure. I've

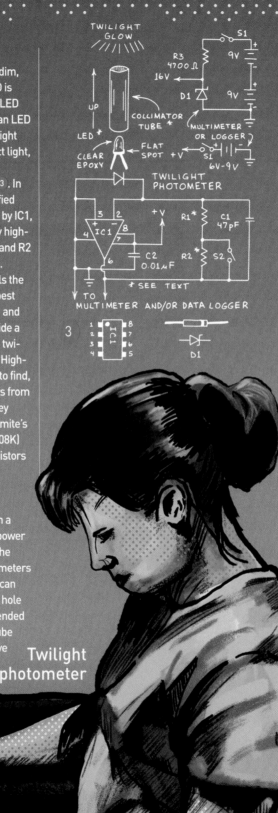

Twilight photometer

Sun below horizon

70,000+ ft.

FORREST M. MIMS III (forrestmims.org), an amateur scientist and Rolex Award winner, was named by *Discover* magazine as one of the "50 Best Brains in Science." His books have sold more than 7 million copies.

Materials

- **Capacitors, ceramic: 47pF (1) and 0.01µF (1)** designated C1 and C2, respectively
- **Zener diode, 16V** D1
- **Operational amplifier (op-amp) IC, TLC271BIP** IC1
- **LED with clear capsule, 660nm red or 880nm near-IR**
- **LED socket (optional)**
- **Resistors: 40GΩ (2), 4.7kΩ (1)** Use the 40-gigohm resistors for R1 and R2 (see text).
- **Switches, miniature SPST** S1 and S2
- **Batteries, 9V**
- **Battery connector, 9V, with wire leads**
- **Battery support bracket (optional)**
- **Audio or phone plug and jack, ⅛"**
- **Output plug and jack** compatible with your data logger
- **Insulated stand-offs (2)** RadioShack #2761381 or similar
- **Perforated circuit board, 1½"×1¼" with copper pads**
- **Metal enclosure**
- **Brass compression union fitting, ⅜"** from hardware store
- **Aluminum or brass tube, about 4" length** from hobby shop; to fit over your LED
- **Bubble level**
- **Hookup wire and hardware**

Tools

- **Soldering iron and solder**
- **Drill and bits**
- **Multimeter**
- **Screwdriver**
- **Data logger (optional)**

PROJECTS
Amateur Scientist

Build a Twilight Photometer

This ultra-sensitive device detects the altitude of dust, smoke, and volcanic emissions high in the sky

Written and photographed by Forrest M. Mims III ■ Illustrated by James Burke

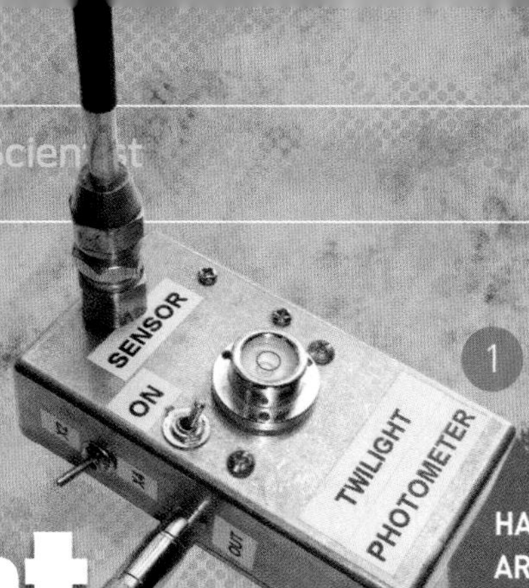

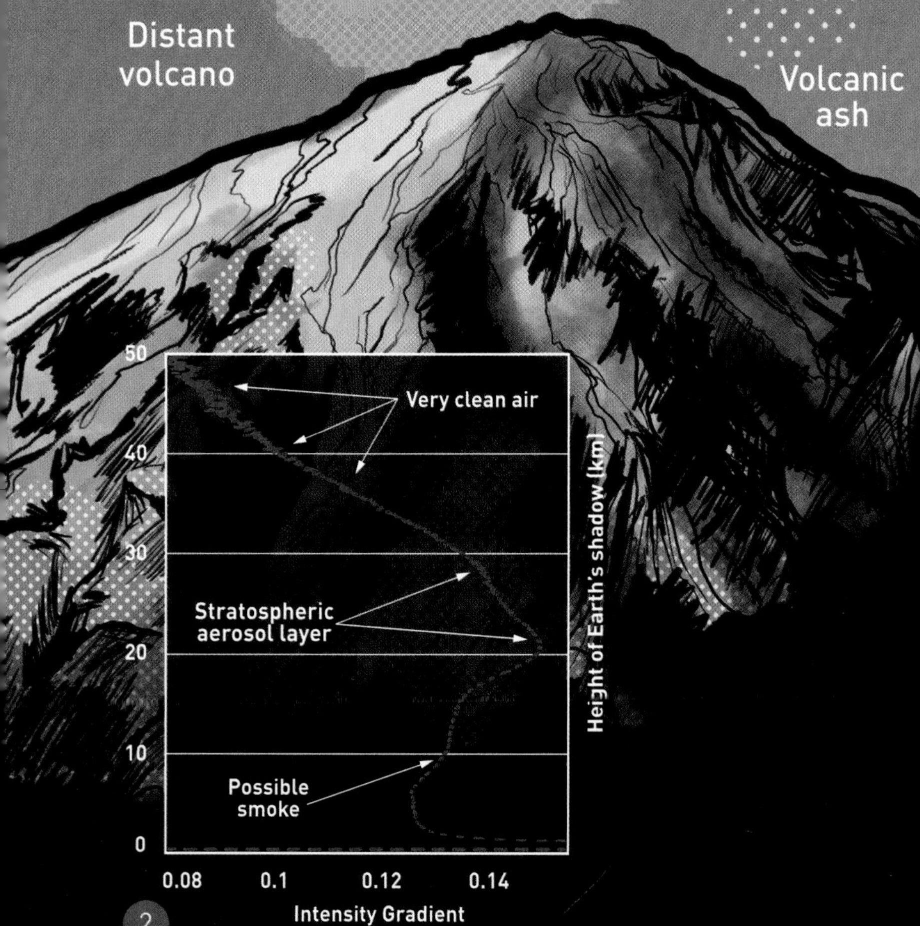

Distant volcano

Volcanic ash

HAVE YOU WONDERED WHY SOME SUNSETS ARE SO SPECTACULAR AND OTHERS SO DRAB? This ultra-sensitive photometer project will allow you to tease out the secrets of twilight and even do serious science by finding the altitude of the dust, smoke, and air pollution that influence the colors of twilight.

With this project you can detect the tiny particles and droplets known as *aerosols* from 3km (around 10,000 feet) high to well above the top of the stratosphere at 50km (165,000 feet). While the photometer will not detect aerosols below 3km, many of those particles eventually float high enough to be detected. For example, from my Texas site I've measured the altitude of smoke from distant fires, haze caused by faraway power plants, and African dust that arrives every summer.

You can even measure the altitude of the sulfuric acid mist that forms an immense blanket 15km–30km high around our entire planet. This stratospheric aerosol layer, which major volcano eruptions can significantly enhance for several years or more, controls the duration of twilights and even influences climate.

BUILD A SIMPLE TWILIGHT PHOTOMETER

The twilight photometer shown in Figure ① requires no optics and is considerably simpler, smaller, and cheaper than those used by professional scientists. Yet, as shown in Figure ②, it nicely estimates the altitude of dust and smoke clouds from 3km to 15km in the troposphere and the permanent aerosol layer at around 15km–30km in the stratosphere. Instead of a conventional photodiode, the twilight glow is detected by an ordinary 660nm red LED or an 880nm near-infrared LED like those used in remote controls for TVs and appliances.

Very clean air

Stratospheric aerosol layer

Possible smoke

Height of Earth's shadow (km)

Intensity Gradient

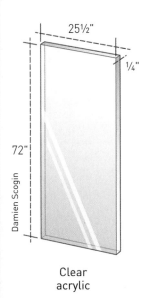
Clear acrylic

One-way mirror window film

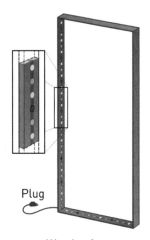

Wooden frame with LED Christmas lights on inside edge

Acrylic mirror

Backing (optional)

3. Screw the 4 boards together to make a rectangular frame 72"×25½", using two 1½" flat-head screws in each corner. Make sure to screw the 24" boards inside the 72" boards.

4. Locate the end of the LED wire that does not have the electrical plug, and use the staple gun to staple it inside the frame at the corner where the filed board is. Now staple the LED lights around the inside of the frame along the midpoint of the 1×4s, with bulbs facing the center of the frame. Staple all the way around the frame.

Partly unscrew the corner with the filed hole and thread the electrical plug through to the outside of the frame. Then rescrew the corner.

5. Drill 12 evenly spaced holes through the clear acrylic sheet, aligned with your frame. Then countersink them.

6. Apply the mirror window film very carefully to the countersunk side of the clear acrylic, following the manufacturer's instructions. Make sure there are no air bubbles. You've just made a very large one-way mirror.

Use the hobby knife to cut out the window film overlapping the countersunk holes.

7. Drill 12 evenly spaced holes through the ordinary acrylic mirror, and countersink these on the backing side (not the mirror side).

Attach this mirror to the frame using 12 screws, with the mirrored surface facing inside.

8. Attach your one-way mirror to the opposite side of frame, with the window film facing inside.

9. Wrap any extra LEDs on the outside of the frame in black tape, or buy a black project box and fold the extra lights into it to hide them.

10. Plug in and enjoy!

USE IT

Hang your mega infinity mirror on the wall as a full-length mirror or on a door to create a life-size portal to another dimension.

Or use it as a tabletop — for a cool coffee table, or a not-at-all distracting beer-pong table! Make sure to use 6 legs and attach them with 2 standard brackets on each leg. Use brackets at least 8" long or else the table will be a little wobbly.

For a beer-pong table, use legs 23½" long, as regulation table height is 27½". (Note that regulation tabletops typically measure 8'×2', where ours is about 6'×2'.) That's a "make"! ◉

See step-by-step photos and share your build at makezine.com/go/mega-infinity-mirror.

**Time Required:
1–2 Days
Cost:
$150–$200**

Materials

» **Dimensional lumber, 1×4, 8' lengths (2)** Nominal 1×4s actually measure ¾"×3½".
» **Spray paint, matte black**
» **Mirror, acrylic, ¼"×72"×25½"** from a plastics store such as TAP Plastics
» **Acrylic sheet, clear, ¼"×72"×25½"** from a plastics store
» **One-way mirror privacy window film** I used Gila brand Privacy Control Mirror window film
» **Wood screws, flat-head, #10: 1½" (8) and ¾" (24)**
» **LED Christmas lights, 100 count**

Tools

» **Drill and bits: 3/16" and countersink**
» **Staple gun**
» **Screwdriver**
» **Round file**
» **Jigsaw or chop saw or handsaw**
» **Hobby knife**

PROJECTS

Easy Mega Infinity Mirror

Written by Lila Becker

This quick build makes a door-sized star portal — or a vertiginous beer-pong table!

LILA BECKER is a designer at SmartLab Toys, with a degree in product design from the University of Oregon. She has always loved building things.

I WAS INSPIRED TO BUILD THIS PROJECT AFTER VISITING A MUSEUM IN SCOTLAND that had a room full of infinity mirrors that made it look like you were floating in space surrounded by stars. I thought it was the coolest thing ever!

My infinity mirror design is definitely a DIY project. Everything you need is easily obtainable at the hardware store and plastics store. The trick is making your own big one-way mirror out of acrylic sheets and privacy window film.

This project is easy to build — no expertise in electronics is needed. You'll use inexpensive electronically controlled Christmas lights, which are available year-round on sites like Amazon. Customize your mirror by choosing lights in your favorite colors or ones that blink or change hues.

1. Cut the 1×4 boards into two 24" lengths and two 72" lengths.

2. Spray-paint the boards matte black. Use a round file to cut a groove about 1" from the end of one 24" board to accommodate the cord from the LED lights.

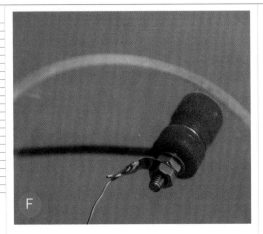

4. ATTACH THE ELECTRODES

To make the electrodes, solder a length of 22–20 gauge solid wire to the solder tab of the binding post terminal. If your terminal doesn't have a solder tab, simply wrap the wire around the inside binding post screw.

Secure each binding post with a nut, then attach the high-voltage wires from the transformer (Figure F).

5. PREPARE THE CIRCUIT FOR AUDIO INPUT

By turning potentiometer R3 with a small screwdriver (Figure G), you can fine-tune the bias for the audio signal to the NPN transistor amplifier. Adjust potentiometer R3 to its midpoint position before turning on the plasma arc speaker.

The audio input signal I used was 100mV (0.1V) peak to peak from an iPod. If the audio signal is too large it will be clipped, creating distortion in the plasma speaker's output. If your plasma arc speaker sounds terrible, the first thing to do is to reduce the volume of your audio signal going into the circuit. Less may be more.

6. ROCK SOME TUNES!

Before you turn on your plasma arc speaker for the first time, adjust the electrodes so the wires are arced and the ends of each wire face each other (Figure H). This is to ensure that the electric arc forms between the wire ends, which will provide the best audio quality. If the arc travels up and down the side of the wire, you'll hear distortion. Adjust the gap between the wire ends to approximately ¼". Then connect your audio player to the circuit's audio input, and press Play (Figure I).

Power up the speaker, and adjust the volume as necessary to form an arc (Figure J). If an arc doesn't form, make sure audio is being played. If the arc still doesn't form, switch the speaker off, unplug it, and then adjust the gap as necessary.

With your plasma arc speaker, you'll be rocking tunes with the same tech-nology that high-dollar audiophile speakers use!

FURTHER EXPERIMENTS

» **Crossover** To use your plasma arc speaker as a tweeter to complement your existing hi-fi speakers, try the bass crossover circuits used by Paul Faget (github.com/paulfaget/PlasmaArcSpeaker) or Oliver Hunt (hvlabs.com/plasmasonic.html).

» **Flame Speaker** You can even use a high-temperature flame (low-temperature plasma) as the gaseous membrane to generate sound. Learn more on the project page online! ◐

✢ More high voltage makers: Richard Hull and the High Energy Amateur Science group do great things with Tesla coils and fusors at tfcbooks.com/mall/hull.htm and teslauniverse.com/community/groups/heas-tcbor

See the plasma arc speaker in action, try more experiments, and share your build on the project page at makezine.com/go/plasma-arc-speaker

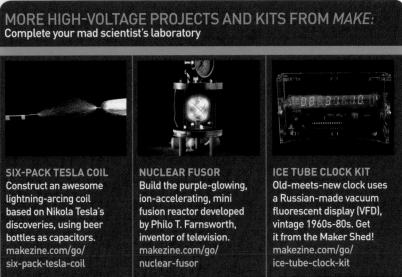

MORE HIGH-VOLTAGE PROJECTS AND KITS FROM MAKE:
Complete your mad scientist's laboratory

SIX-PACK TESLA COIL Construct an awesome lightning-arcing coil based on Nikola Tesla's discoveries, using beer bottles as capacitors. makezine.com/go/six-pack-tesla-coil

NUCLEAR FUSOR Build the purple-glowing, ion-accelerating, mini fusion reactor developed by Philo T. Farnsworth, inventor of television. makezine.com/go/nuclear-fusor

ICE TUBE CLOCK KIT Old-meets-new clock uses a Russian-made vacuum fluorescent display (VFD), vintage 1960s-80s. Get it from the Maker Shed! makezine.com/go/ice-tube-clock-kit

PROJECTS | Plasma Arc Speaker

CAUTION:
HIGH VOLTAGE CAN BE LETHAL

If you're not familiar with working with high-voltage devices, we do not recommend you build this device. If you have a weak heart or an implanted biomedical device like a pacemaker, do NOT build this device. Safe assembly and operation of this project is the user's responsibility. A high voltage may cause you to jump, move, or fall, and can thereby cause a secondary injury unrelated to the electric shock itself. Take the following precautions and treat all high-voltage power supplies with the respect they deserve.

BASIC SAFETY GUIDELINES

» Keep one hand in your pocket. Only use your other hand to work with the high-voltage equipment. This reduces the probability of accidentally passing high-voltage current across your heart from hand to hand.

» Set up your work area away from possible grounds that you may accidentally contact. Keep your work area neat and clean to easily identify high-voltage wires and grounds.

» Be sure the floor is dry and wear rubber-soled shoes.

» Prove to yourself the high-voltage power supply is off — by unplugging the device's electrical power cord. Don't trust power switches that could be accidentally turned on.

» Do not work on high-voltage apparatus when you're tired or not alert.

» High voltage could jump to the low-voltage side of the circuit and damage the audio player. Use only inexpensive audio devices. We are not responsible for any damage to your audio devices.

For more safety guidelines, visit the project page at makezine.com/go/plasma-arc-speaker

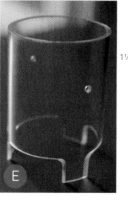

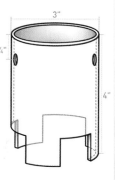

1. BUILD THE CIRCUIT

Solder the components to the printed circuit board (PCB) included in the kit (Figure B). If you're not using the kit, you can prototype the circuit using point-to-point wiring, following the schematic (Figure A). You can use a solderless breadboard for the *low-current components only*, like the 555 timer, 2N3904 transistor, and audio input. (While the circuit as a whole draws less than 2A of current, that's still above the rating of solderless breadboards.)

NOTE: If you're using a different HV transformer, you may need to adjust the base frequency of the 555 timer in order to achieve a "silent" high-voltage plasma arc. Experiment with the resistor and capacitor values of the RC network (R5, R6, and C3) to tune the base frequency until a silent electrical arc is achieved.

Spread heat sink compound on the back of the IGBT and make sure there's good contact with the heat sink to ensure good heat transfer (Figure C).

2. MOUNT IT IN THE ENCLOSURE

The output of this circuit generates high voltage — be careful near the output of the HV transformer! — so you'll mount it in a plastic enclosure, as plastic is an insulator and will be safer than a metal enclosure. First, connect the high voltage wires to the output of the HV transformer; you will run these out to the binding posts of the plasma speaker tube.

Drill and cut holes in the enclosure for the fan venting, power jack, audio input, LED, switch, and high voltage wires (Figure D), following the template on the project page at makezine.com/go/plasma-arc-speaker.

IMPORTANT: The large IGBT heat sink and 12VDC blower fan are essential. Mount the fan close to the heat sink to maximize cooling, and make sure your enclosure has large circular cutouts for intake and exhaust to permit sufficient airflow for cooling. Without taking these actions, your IGBT will overheat in less than a minute.

3. BUILD THE PLASMA SPEAKER TUBE

The speaker is fabricated out of a 4" length of 3"-diameter clear plastic tubing. Cut 3 notches in the bottom of the plastic tube to create legs and to allow airflow. (The arc can get as hot as a candle.)

Drill holes on opposite sides of the tube, 1¼" from the top, sized to accept the binding post terminals (Figure E).

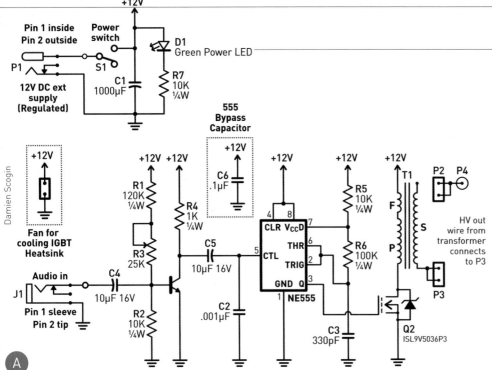

Time Required: 2-5 Hours
Cost: $50-$100

Materials

- **Plasma Arc Speaker Kit** includes all parts listed below, $90 from Images SI Inc., imagesco.com/kits/plasma-speaker.html
- **Heat sink compound**

— Or buy the following parts separately —

- **High voltage flyback transformer, 40W–80W, 20kV max output** available separately from Images SI, part #HVT-01. Designated T1 in the schematic.
- **Power jack, 2.1mm** P1
- **Toggle switch** S1
- **Resistors, 100kΩ, ¼W (2)** R1, R6
- **Resistors, 10kΩ, ¼W (3)** R2, R5, R7
- **Resistor, 1kΩ, ¼W** R4
- **Potentiometer, multi-turn trimmer, 25kΩ** R3, can use 10kΩ–25kΩ
- **Capacitor, 470µF, 25V** C1, can use 470µF–1,000µF, 16V or higher
- **Capacitor, 0.001µF, 100V** C2
- **Capacitor, 330pF, 50V** C3
- **Capacitors, 10µF, 16V (2)** C4, C5
- **Capacitor, 0.1µF, 100V** C6
- **LED, submini, green** D1
- **Audio jack, 3.5mm** J1
- **555 timer IC chip, LM555** U1
- **IC socket, 8-pin**
- **Transistor, 2N3904** Q1
- **Transistor, IGBT, ISL9V5036P3** Q2, also available separately from Images SI
- **Large heat sink for IGBT**
- **Fan, 12VDC, 40mm×20mm**
- **Plastic tube, 3" diameter, 4" length**
- **Binding post, black**
- **Binding post, red**
- **High voltage wire, 12"–18" length**
- **Wire, 22–20 gauge, solid insulated, 12" length**
- **Power supply, 12VDC, 2A or greater**
- **Plastic project enclosure**

Tools

- Soldering iron
- Wrench or pliers
- Rotary cutting tool
- Drill and drill bits
- Breadboard (optional) if you're not using the custom PCB from the kit

WHAT DO THE WORLD'S MOST EXPENSIVE SPEAKERS, THE SUN, AND LAMPS FROM THE 1890s HAVE IN COMMON? Plasma, the fourth state of matter, of course! After building this little plasma arc speaker, you'll be able to hear your favorite jams played directly from the vibrating plasma of an electric arc.

Plasma is a high-temperature, highly ionized gas that's electrically conductive. While conventional loudspeakers use a solid diaphragm, the plasma arc speaker uses ionized gas as a gaseous diaphragm — it's virtually massless, so it easily responds to high-frequency audio signals. By varying the electrical signal across the electrodes, you'll cause the ions in the plasma to jiggle, which causes the gaseous diaphragm to vibrate and create sound waves in the air.

THE SINGING ARC

The first instance of a plasma arc speaker can be traced back to William Duddell in 1899. Duddell connected an ordinary carbon arc lamp to a tuned circuit made of a capacitor and inductor, and he discovered that he could generate tones that corresponded to the resonant frequency of the tuned circuit. He wired a keyboard and played "God Save the Queen" on what is considered the first electronic musical instrument — the "singing arc."

HOW IT WORKS

Most modern published schematics for building "singing arc" speakers use either a 555 timer or a TL494 PWM controller, whose output is wired to a standard power transistor or MOSFET to rapidly switch current on and off to the high-voltage (HV) transformer (Figure A). I chose to use a 555 timer with a unique power component, the insulated-gate bipolar transistor (IGBT), which is ideally suited for this application. It's got the high current capacity of a bipolar transistor and the voltage control of a MOSFET.

The 555 timer is set up in astable mode to continually output a frequency dependent upon an RC network composed of two resistors and a capacitor. This is the oscillation frequency that powers the HV transformer. I chose a "base frequency" of approximately 23kHz because it allows the HV transformer to create an output arc (plasma) that doesn't produce any sound or tone on its own, as this would detract from the sound quality of the speaker.

Pin 5 on the 555 timer is the control voltage input. By applying a voltage to this pin, we can vary the output frequency of the timer independently from the base frequency that's set by the RC network. This creates a frequency modulation (FM) output, like FM radio. Just connect the audio output from the 2N3904 transistor to pin 5, and now your audio signal is modulating the output frequency of the 555 timer. This FM output, amplified by the HV transformer, is what jiggles the ions in the plasma arc to create sound.

This small speaker is the equivalent of a tweeter. Large electric arcs can produce better fidelity at lower frequencies, but the smaller arcs in this device are better at reproducing sound in the higher frequencies.

Plasma Arc Speaker

Written by John Iovine

Create a high-voltage loudspeaker that makes sound waves by vibrating an electric arc

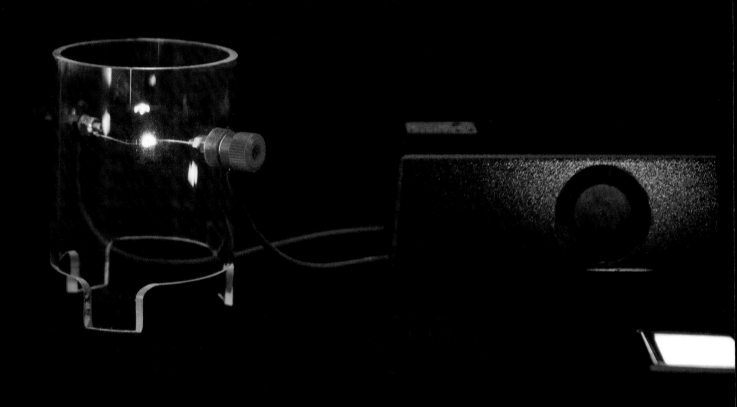

JOHN IOVINE is a science and electronics tinkerer and author who owns and operates Images SI Inc., a small science company. He resides in Staten Island, N.Y., with his wife and two children, their dog, Chansey, and their cat, Squeaks.

Use an angle grinder or Dremel to cut the neck down lower if you wish.

Now measure the thickness of your soundboard (mine was 2.6mm), and cut the body down lower than the neck by that exact thickness, so the soundboard will lie flush beneath the fretboard. I marked the cut line with painter's tape (Figure O and P).

9. INSTALL THE GUITAR COMPONENTS

Installing the wood components is exactly like making a traditional guitar. I glued them to the body with 5-minute epoxy, using clamps and weights: first the soundboard, then the headstock, and finally the fretboard and bridge. (Figures Q, R, and S).

You'll need to cut the neck to fit your headstock, then cut your fretboard to fit the neck; for details, see the project page at makezine.com/go/carbon-fiber-acoustic-guitar. I also recommend *Build Your Own Acoustic Guitar* by Jonathan Kinkead.

10. PATCH, SAND, AND POLISH

Patch any gaps with black 5-minute epoxy or your layup epoxy, tinted black (Figure T).

Protect the wooden components with masking tape and sand down any drips or high spots with 80 or 120 grit. Then wet-sand with waterproof sandpapers in 180, 220, 320, and finally 400 grit (Figure U).

At this point I gave the whole body another gloss coat (optional) of epoxy (Figure V), and let it dry a full day, then wet-sanded with 400, 600, and 1000 grits to get a semi-gloss shine.

To polish the body, I gave it a coat of Turtle Wax auto-body rubbing compound, then a final coat of paste wax (Figure W).

This project pushed me to learn a lot of new skills and try experiments I never would have attempted otherwise. The fretboard is inlaid two ways: with CNC-cut maple and with laser-cut lettering filled with colored epoxy; the headstock is also laser-inlaid and the bridge is CNC-cut too. Finally, I laser-etched the Western red cedar soundboard with my own artwork, then hand-tinted it (Figure X).

Considering this was my first time working with carbon fiber — or making a musical instrument — I'm very happy with how it all turned out. ⬤

Get the 3D guitar body model and more tips and photos at makezine.com/go/carbon-fiber-acoustic-guitar.

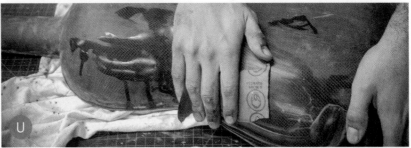

PROJECTS | Carbon-Fiber Acoustic Guitar

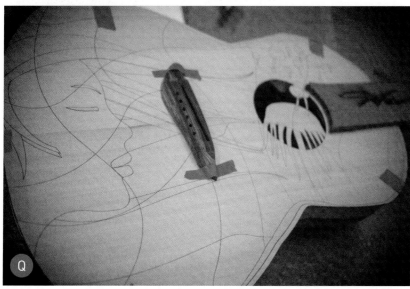

7. REINFORCE THE BODY

A guitar body must be sturdy — any energy spent flexing is energy taken away from the sound — and the neck must withstand more than 100lbs of pull from the strings.

Check for any areas that are flexing too much. The edges of my guitar flexed a little bit, and so did some parts of the neck. (A traditional guitar has a threaded truss rod to adjust stiffness, but ours just has a very stiff carbon neck.)

Coremat is a super absorbent foam that sucks up lots of epoxy, making it very stiff once it dries. I put a layer of Coremat all along the edges of the body and down the center from the neck, adding a strip of S-Glass under the Coremat to help it stick to the carbon. It's not pretty, but nobody will see it anyway (Figure M).

Flip the guitar upside down on a plastic mat and let the epoxy drip down to form a "lip" around the inside edge of the guitar. This makes a nice surface for mounting wooden components later. Weight the body and neck so they'll dry flush (Figure N).

8. FLATTEN AND SHAPE THE BODY

Check that the guitar lies flat, and sand down any high spots with 80- or 120-grit paper on a long sanding block made of scrap 2 × 4.

Check the size of the neck and how it feels in your hands. Compare it to other guitars you play.

type of barrier. Commercial mold release products won't work with cardboard, so you'll wrap the entire mold in packing plastic — a thicker, tougher cousin of cling-wrap (Figures H and I). Aim for long, smooth motions to avoid tangling and bunching.

5. LAY UP THE CARBON FIBER

Protect your workspace with tarps or plastic wrap, then neatly arrange your materials before starting. Once you mix the epoxy, things get messy fast!

Cut 2 sheets of carbon fiber and 1 sheet of S-Glass just a little longer and wider than the mold (I had about ½" overlap on each side). Set aside the better-looking sheet of carbon; this will be your final, show layer.

Mix 10oz of epoxy following the manufacturer's instructions, in this case 4 parts resin to 1 part hardener. You can eyeball it in a measuring cup, but it's best to measure it by weight. Epoxy likes it warm, so be aware of your epoxy's minimum recommended temperature.

Lay the first layer of carbon over the mold, then load up your brush and wet the carbon fiber thoroughly with epoxy — no dry spots or bubbles (Figures J and K). Your epoxy has a 20–30 minute "pot time" before it hardens.

> **TIP:** In tight corners the carbon fiber will tent and pull away from the mold. Cut relief slits, then overlap the 2 sides. If needed, bulk up the area with scraps of carbon fiber to soften the curve. Don't worry about how it looks (it is only the first layer) and don't be afraid to use your (gloved) fingers to tug, smoosh, and reshape the fiber by hand.

Mix another batch of epoxy and put down a layer of S-Glass in exactly the same way. Then another batch and your final layer of carbon. Careful, as everything you do now will show on the surface.

Let the epoxy cure fully, protecting it from dust and fingerprints.

6. RELEASE THE MOLD

Wearing your goggles and respirator, use a Dremel, angle grinder, or hacksaw to cut away the excess carbon and epoxy fringe to just above the cardboard mold. Beware of sharp little spikes and splinters.

Have a friend hold the guitar body while you tug the mold out, or just tear it out in pieces (Figure L). Examine your handiwork. The hardest part is over!

PROJECTS | Carbon-Fiber Acoustic Guitar

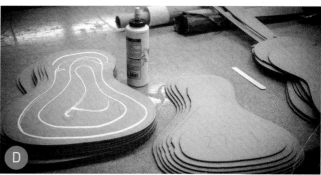

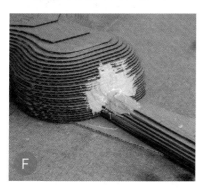

3. LASER-CUT AND ASSEMBLE THE MOLD

I used Adobe Illustrator to send the EPS files to an Epilog 60W laser cutter (Figure C). Choose settings appropriate for cardboard (I used 70% power, 90% speed, and 500Hz frequency).

Each part is numbered. Stack them in numerical order, then assemble with a little glue between each layer (Figures D and E). Clamping is optional.

4. PREPARE THE MOLD

» Soften edges. Carbon fiber doesn't bend well over sharp corners. Use auto-body putty to smooth the joint between the body and neck (Figure F).

» Make a handle. Nail boards to the bottom of the mold to make a firm handle. Then clamp it in a vise to elevate the mold and keep it rock steady (Figure G).

» Mold release. To prevent the epoxy from bonding with the cardboard, you need some

makezine.com/44

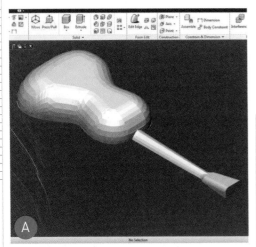

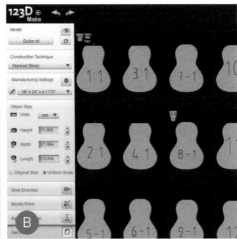

HERE'S HOW I MADE THE BODY FOR MY CARBON-FIBER ACOUSTIC GUITAR, AS PART OF A WORKSHOP AT TECHSHOP in San Francisco. It's a great technique that anyone can do. First you design a 3D mold made of stacked-up slices of cardboard, using Autodesk's free 123D Make software. Then you lay carbon-fiber fabric over the mold and harden the fibers with marine-grade epoxy. My mold is modeled after a 1940 Gibson L-00 but the sides are rounded like an Ovation-style guitar, because carbon fiber doesn't bend well around sharp corners. The finished guitar looks and sounds great!

I designed the workshop for complete beginners, so the whole body is laid up and finished with brush and gloved hands. (A more advanced technique uses vacuum bags, which I hope to try in the future.) It took me a week, working nights, to finish the body alone. You can probably make the whole guitar in a couple of weekends.

1. GET THE 3D GUITAR MODEL
You can design your own guitar using your favorite 3D design software, or skip it and just download the 3D model I used (Figure A). Get it at the project page online: makezine.com/go/carbon-fiber-acoustic-guitar.

2. SLICE THE MODEL INTO LAYERS
Open Autodesk 123D Make and import your 3D model. On the left, under Construction Techniques, select Stacked Slices.

Now set the dimensions for the object you're making and the material it's being made from. For the Object Size, my model was 935mm long × 358mm wide × 111mm high (36"×14"×4.37"). For my Manufacturing Settings, I chose 0.177" slices on 18"×24" sheets, which created 56 parts spread over 23 sheets of standard cardboard (Figure B).

Inspect the model and sheets for errors. Then click Get Plans and select EPS format.

Time Required: A Couple Weekends
Cost: $180–$200 body $100 components

Materials
- **Corrugated cardboard**
- **Carbon-fiber fabric** I used plain weave because it was affordable, but other styles might work better.
- **Fiberglass, S-Glass (optional)** A cheaper filler used between carbon layers.
- **Coremat polyester foam**
- **Epoxy, TAP Plastics Marine Grade Premium 314 resin and 109 hardener** Long open time, low odor, few bubbles.
- **Black pigment** for tinting plastic resins
- **Plastic packing wrap** aka stretch wrap or pallet wrap
- **Auto body putty** aka Bondo
- **Painter's tape**
- **Scrap wood**
- **Sandpaper,** 80 or 120 grit
- **Sandpaper, waterproof:** 180, 220, 320, 400, 600, and 1000 grits
- **Auto-body rubbing compound**
- **Paste wax**

Tools
- **Computer**
- **3D design software (optional)** My instructors used Autodesk Inventor to design the guitar's 3D model. Download it from the project page, makezine.com/go/carbon-fiber-acoustic-guitar.
- **123D Make software** free from 123dapp.com/make
- **Laser cutter (optional)** You can cut the cardboard by hand but I recommend you visit makezine.com/where-to-get-digital-fabrication-tool-access to find a machine or service you can use.
- **Vector graphics software (optional)** such as Adobe Illustrator or CorelDRAW, to send files to the laser cutter
- **Vise**
- **Scale (optional)**
- **Disposable containers and brushes**
- **Disposable gloves**
- **Dust mask or respirator** Carbon fiber and epoxy make some nasty dust when cut and sanded.
- **Safety goggles**
- **Disposable scissors**
- **Hacksaw**
- **Rotary tool with cutoff wheel** such as a Dremel
- **Sanding block**
- **Rag or polishing pad**

WHY CARBON FIBER?
Carbon fiber won't replace traditional wood-body guitars anytime soon, but it has definite advantages:

- **Acoustics** Good carbon-fiber guitars have a very clear sound and loud volume. And carbon fiber varies little compared to wooden components, so it's easier to crank out duplicates with reliable sound qualities.
- **Durability** Carbon fiber is just plain durable and strong. It's a lot easier to patch or rebuild than wood, and it's not affected by weather and humidity.
- **Ease of Construction** In just one night I went from a bottle of epoxy and a pile of fabric to a nice guitar-shaped body ready for finishing. Customization is easier too. You don't need special guitar jigs, clamps, and tools — just reshape your mold.

PROJECTS

Carbon-Fiber Acoustic Guitar

Written by Seth Newsome

Use free 3D design tools to mold a tough, sleek guitar body (or anything else) in high-tech composite

SETH NEWSOME is a teacher turned maker living in San Francisco. He loves to explore how to create things (especially in wood), to master new methods, and then pass on his knowledge and teach others how to do the same thing.

lens from below and the other from above. This increased my magnification to ~325x.

Be careful not to let these lenses touch, and install them as level as possible. Failure to do so can cause aberrations in the image.

5. PLACE THE LIGHT SOURCE
Next, drill a shallow hole in the base to accommodate your light source (Figure G). (Small LED click lights work well.) It's important that your light source is directly below the focus lens. To mark the right placement, first test-fit the base, bolts, and camera stage together, then slide the camera stage down to the base and mark the base with a pencil, directly beneath the lens.

6. ASSEMBLE THE MICROSCOPE
Start with washers and nuts to hold the 3 bolts tight to the base. Add some upside-down wing nuts and then washers to the 2 front bolts. Then place the specimen stage on top of these washers (Figure H).

Add a nut to each bolt, and lower them about ½". Then put a compression spring over each front nut and bolt, and rest the camera stage on top of the nuts (Figure I).

A level is handy here to make sure that the stage is actually flat. If you don't own a level there are plenty of free level apps for a phone! Use wire cutters or pliers to trim or adjust the springs if necessary. When the stage is level both front to back and left to right, tighten down the final nuts.

The compression springs keep the specimen stage stabilized and allow you to make far finer adjustments with the wing nuts. Without these springs, the specimen stage can tilt one way or another if the load is imbalanced or if the bolt holes are too large.

7. MAKE SOME SLIDES
The focal length of the lens is very short and the specimen stage can only be raised so close to it because of the nuts holding up the camera stage. Using a transparent sample slide fixes this problem and makes manipulating samples while viewing easier. Simply cut a 2"×4" piece of plexiglass.

With 2 lenses, the focal length gets even smaller, so 2 plexi slides are required.

8. TAKE AMAZING PICTURES!
Or video! With $10 worth of materials and a smartphone, you just made a digital microscope.

Align your smartphone camera lens with the microscope lens. Bring the object into focus by slowly turning the wing nuts on either side, then use your phone to take a picture or video, or even zoom in for a closer look.

I'm a major proponent of making home science more accessible. My goal in designing and building this phone-to-microscope conversion stand is to provide an alternative to overly expensive microscopes. This setup is a viable option for underfunded science classrooms that would not otherwise be able to perform experiments requiring a microscope. But more than that, this device will allow people to rediscover the world around them.

Cricket wing

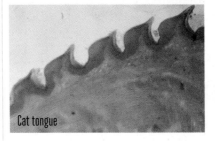
Cat tongue

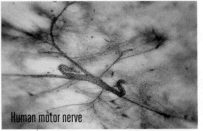
Human motor nerve

See more step-by-step photos and micro photos and share your own at makezine.com/go/smartphone-microscope.

+SKILL BUILDER
DRILL ACRYLIC WITHOUT CRACKING

Acrylic is prone to cracking when cut or drilled. Here's how to prevent it: First, put a piece of tape over the area you're about to drill. Measure and mark, then drill *slowly*, using a sharp bit and gentle pressure. Do not press hard — let the drill do the work.

PROJECTS | Smartphone Microscope

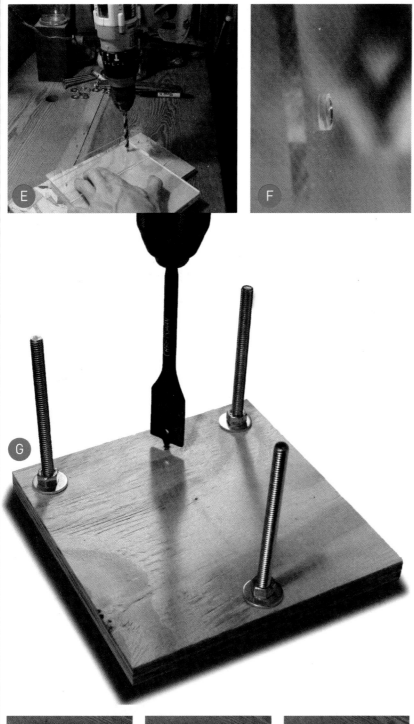

2. CUT AND DRILL THE STAND

Mark the top of the plywood base at the front 2 corners, ¾" in from both the sides and the front edge. Make a third mark centered ¾" from the back edge.

Stack the plexiglass camera stage (7"×7") on top of the base. Then stack the specimen stage (3"×7") on top of the camera stage, with ¾" of the specimen stage extending over the front of the base. Clamp and drill through the entire assembly (Figure E).

TIPS: Put a sacrificial piece of wood beneath the plywood base before drilling. You don't want to damage your workbench.
To avoid cracking the acrylic, see "Skill Builder" on the next page.

The bolts that stick up through the base must be counterbored in order for the stand to sit flat. Flip the base over and counterbore the holes with a spade bit to accommodate the bolt heads.

3. MOUNT THE LASER LENS

Find a drill bit slightly smaller than the diameter of the lens. Then measure, mark, and drill a hole in the camera stage, ¾" from the front edge, in line with the bolt holes.

Press-fit the lens into the hole (Figure F). If it doesn't quite fit, use a file or sandpaper to enlarge the hole. Work slowly and test the fit often. It's easy to overshoot and make the hole too large! You can remedy this with tiny bit of glue, but be very careful not to get glue on the lens surfaces.

When using the stand, it's important to have the lens as close as possible to the camera. If you plan to keep your phone in a case when you use the stand, then leave the top of the lens slightly exposed so that it will rest closer to the camera. Otherwise, mount the lens flush with the stage.

4. MOUNT SECOND LENS (optional)

If you're using 2 laser lenses, mount them one above the other in the camera stage, inserting one

 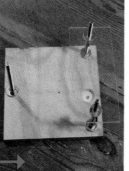 

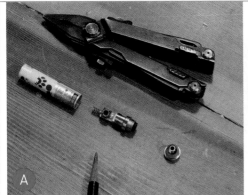

KENJI YOSHINO studied chemistry at Grinnell College and currently works for Grin City Collective, an art residency in central Iowa.

Time Required:
20–30 Minutes
Cost:
$10–$15

Materials

- » **Your smartphone**
- » **Laser pointer focus lens (1 or 2)** salvaged from $2 laser pointers, or bought separately at store.laserclassroom.com/laser-pointer-lens
- » **Plywood, ¾"×7"×7"**
- » **Acrylic sheet, ⅛": 7"×7" (1), 3"×7" (1), and 2"×4" (1 or more)** aka plexiglass. Ask your hardware store to cut it. If you're cutting your own, ask if they'll sell you scraps to save money.
- » **Carriage bolts, 5/16"×4½" (3)**
- » **Nuts, 5/16" (9)**
  Wing nuts, 5/16" (2)
- » **Washers, 5/16" (5)**
- » **Compression springs, ~20 gauge, ¾" ID, ½" long (2)** easily compressed with your fingers. You can buy a longer one and cut it into shorter pieces.
- » **Light source (optional)** for viewing backlit specimens. I like cheap LED click lights or stubby LED flashlights like Diamond Visions #08-0775.

Tools

- » **Drill and drill bits**
- » **Clamp**
- » **Saw(s)** for cutting plywood and acrylic (if your hardware store won't)
- » **Ruler**
- » **File**
- » **Sandpaper**
- » **Wire cutters**
- » **Pliers**
- » **Level**

YOU CAN EASILY TURN YOUR SMARTPHONE CAMERA INTO A POWERFUL DIGITAL MICROSCOPE. All you need is a few tools, the focus lens from a cheap laser pointer or two, and about $10 worth of materials from the hardware store.

Not only will this homemade microscopy stand take high-quality macro photos, but with the ability to magnify objects up to 175x (or 325x if you use two lenses), you can easily see and photograph cells. You can even do laboratory experiments — we were able to observe plasmolysis in red onion epidermal cells.

You're not restricted to a laboratory setting with this microscope. It was designed to be easy to operate, lightweight, and portable. Just align your phone's camera with the focus lens on top of the camera stage, then place the object you'd like to view on the adjustable specimen stage.

Since I first shared this project on Instructables, I've added a second lens for higher magnification, springs to keep the specimen stage steady, and plexiglass slides to make switching samples clean and easy.

Because the stages are also constructed with plexiglass, objects can easily be viewed with or without an external light source. This lets you use the microscope in a wide range of settings — in the classroom, outside, or in your own home — to take a closer look at the world around you.

Here's how to make it.

1. EXTRACT THE LASER LENS (optional)

The focus lens of just about any laser pointer will serve as the macro lens on the microscope stand. Don't waste money on an expensive model; the lens from a $2 laser is fine. To achieve higher magnification (up to 325x), use two! You can also buy the bare lenses online and skip this step.

To get the lens from the laser pointer, start by unscrewing the front cone and the back cover of the tube. Remove the batteries. Using the eraser end of a pencil, push the innards out of the front of the tube (Figure **A**).

The front of this assembly is where the focus lens sits. Unscrew the small black plastic retainer in front of the lens and the lens will come free (Figure **B**).

The lens is not symmetrical when viewed from the edge. You'll see a thin, translucent strip (~1mm) on one side of the otherwise transparent lens (in Figure **C** it's shown on the left side). This side must face away from the camera. You can determine the correct orientation by sticking the lens between the prongs of a hairpin, then taping the rig to the back of your smartphone as shown in Figure **D**. The correct orientation will provide you with a larger field of view.

As it is, you can take reasonably good macro photos with just this lens and your smartphone. But it's extremely hard to keep the phone steady when taking zoomed-in photos. That's why you need to build a stand!

PROJECTS

Smartphone Microscope

Written by Kenji Yoshino

Take amazing high-magnification photos with your phone, using the lens from a cheap laser pointer

FLYING OVER OPEN WATER IS EXCITING, BUT A GOOD DAY of R/C fun can be ruined by even a relatively small puddle. Fortunately, you can protect your electronics investment from a dunk with a few readily available products.

PLANNING YOUR APPROACH

The majority of what we do at Flite Test involves radio-controlled airplanes or multirotors, so the components we want to treat include transmitters, receivers, electronic speed controllers (ESCs), servos, and flight control boards. For most of these, a petroleum-based hydrophobic agent such as CorrosionX applied liberally will be sufficient to protect against incidental exposure to moisture, for example flying in light rain or snow. (Conformal coating, essentially spray enamel, is another option for receivers, but it's not recommended for ESCs or motors because it will clog the bearings.)

In fact, we submerged a small R/C helicopter coated in CorrosionX in fresh and salt water and were able to fly it out of both (Figure A)! Of course, this is an extreme test and not everyone is planning on dunking his or her electronics in a tub of water. If you do anticipate more extensive or prolonged exposure, that might change how much time and effort you're going to spend making everything water resistant.

PROTECTING CRITICAL SYSTEMS

For servos, we recommend CorrosionX's denser HD product; you'll need to open up the housing and spray inside (Figure B). The potentiometer (a variable resistor) is the most critical part to waterproof because if water gets inside it will give you false readings. When you're finished, be sure to clean the outside with isopropyl alcohol or glue won't stick to it.

For more sensitive electronic components such as ESCs we suggest using heat-shrink tubing and epoxy to create a protective barrier. Note: Hot glue will *not* work for this because it doesn't stick to silicone. First remove any manufacturer-applied heat-shrink tubing and replace it with a larger piece. Apply heat as you normally would, then seal the ends with 5-minute epoxy, making sure not to leave any gaps (Figure C). If you want to make the seal waterproof, instead of just resistant, you can use 24-hour epoxy.

A NOTE ON MOTORS

If you're using brushless motors, they are, by their nature, waterproof and will operate while submerged, assuming the wiring is properly insulated. The bearings should be oiled after exposure to water, though.

With a few precautionary measures, you can feel more confident flying over lakes, in inclement weather, or wherever there may be water involved! Just be sure to brush up on your swimming and SCUBA skills.

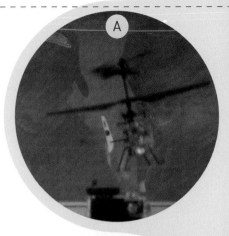

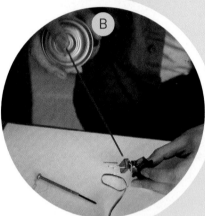

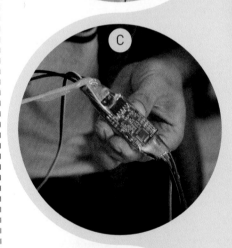

AUSTIN FUREY is the marketing manager for FliteTest.com, a community dedicated to entertaining, educating, and elevating the world of flight.

Time Required: 10–15 Minutes
Cost: $50–$60

Materials
» **CorrosionX** (or conformal coating such as Turbo-Coat)
» **CorrosionX HD**
» **Heat-shrink tubing, clear**
» **Epoxy, 5-minute**
» **Alcohol, isopropyl**

Tools
» Screwdriver to open servo housing
» Heat gun for heat-shrink tubing

SPECIAL SECTION

DRONES | WATERPROOF YOUR DRONE

DON'T BE DEAD IN THE WATER

HOW TO PROTECT YOUR ELECTRONICS FROM THE ELEMENTS

WRITTEN BY AUSTIN FUREY AND THE FLITETEST.COM COMMUNITY

COME ON IN, THE WATER'S FINE!
WATERTIGHT MULTIROTORS FROM QUADH2O
WRITTEN BY GRETA LORGE

IF YOUR FLIGHT PLAN INVOLVES MORE THAN THE OCCASIONAL DOUSING, QUADH2O'S MULTIROTORS — available either as a kit or ready-to-fly — are designed to safely land in the water or fly in the rain. The company got its start in 2012 when designer/fabricator Nick Wadman was asked to photograph real estate using an aerial drone and needed to fly over water. Having a background in R/C and a passion for tinkering, he developed the QuadH2o ($849 kit; $3,499 RTF), designed to house the DJI Naza GPS and to carry a GoPro. The second drone in their fleet, the HexH2o, announced in late 2014 and available for preorder ($895 kit; $3,659 RTF), adds the capability to capture stabilized still or video footage underwater thanks to a watertight gimbal housing built for the DJI Zenmuse and GoPro. Videos of the HexH2o in action show it landing on calm water, maneuvering slowly while filming beneath the surface, and then taking off to capture aerial shots.

SHOOT THE CURL WITH A DIY WAVECOPTER
BUILD A WATERPROOF QUAD FRAME FROM ELECTRICAL PARTS
WRITTEN BY ALEC SHORT

I BUILT THE WAVECOPTER TO GAIN A NEW PERSPECTIVE ON SURFING PHOTOGRAPHY and to do aerial surveying of event sites.

A PVC 4-way junction cut to fit inside a weatherproof electrical socket box forms the hub. The carbon-fiber rotor arms are secured in place with locking PVC conduit thread adapters. The motor mounts are made from PVC 3-way inspection tees with both ends sealed with rubber grommets. (It's important to upgrade the screws that come with the inspection cover to machine bolts, otherwise the motors will pull them free.) Configuration of the power and flight controller electronics will vary depending on your specific setup, but the most important part of this project is being able to house an optimum balance of battery power and weight in a watertight frame. I found that two ZIPPY Flightmax 2200mAh LiPo batteries fit perfectly under the lid of the Diall model WP23L socket box I used, but you may need to experiment a bit with comparable components available near you.

To complete your drone, attach landing gear and floats to the undercarriage. Calibrate your flight controller per the manufacturer's instructions. Then install the motors; as long as they're brushless (most are) no additional waterproofing is necessary, but you can take it further with our moisture-protecting electronics tips on the following page. Connect the batteries and test all your flight control systems before attaching the props to the motors. Happy flying!

Full instructions at makezine.com/go/waterproof-wavecopter/

SPECIAL SECTION

DRONES | NOODLE DRONE

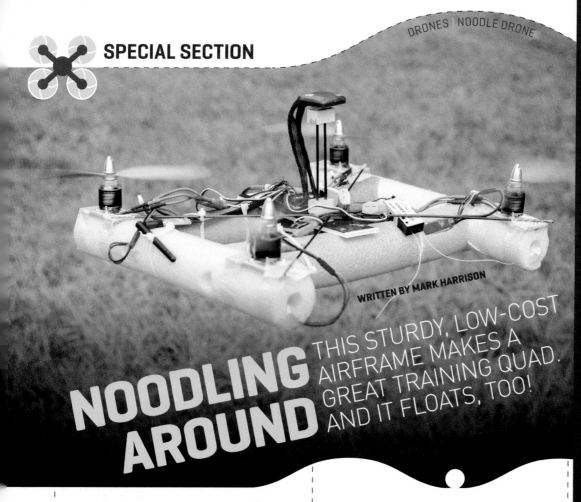

WRITTEN BY MARK HARRISON

NOODLING AROUND

THIS STURDY, LOW-COST AIRFRAME MAKES A GREAT TRAINING QUAD. AND IT FLOATS, TOO!

Time Required:
2 Hours
Cost:
$8

MARK HARRISON is a tech lead at Pixar and an intrepid drone hobbyist who blogs at eastbay-rc.blogspot.com.

Materials
- **Pool noodles** I chose 3 colors so I could easily see the direction the quadcopter was pointing.
- **Motors, brushless outrunner,** Prop Drive 28-30, 900 KV (4)
- **Props, clockwise,** APC 10×4.7 Slow Fly Prop (2)
- **Props,** APC 10×4.7 SFP (2)
- **Electronic speed controllers (ESCs),** 30A (4)
- **Flight control board**
- **Battery,** 3S 2,200mAh
- **Carbon fiber rods,** 3mm: 12" (2) and 15" (2)
- **Foam-Tac glue**
- **Reinforced strapping tape**
- **Velcro straps**
- **Scrap plywood**

Tools
- Hot wire foam cutter
- Hacksaw
- Scissors
- Drill and drill bits
- Screwdriver

LET'S FACE IT: THERE'S JUST SOMETHING AMUSING ABOUT POOL NOODLES IN FLIGHT. I designed and built this copter in response to a *CrashCast* challenge to build a cheap, sturdy, flyable quad. It was bodged together in an evening with materials at hand, but I was happy enough with the result that I use it as a trainer when somebody wants to try flying. (I can't imagine much you could do to break a pool noodle!)

1. CUT AND TRIM THE ARMS
Cut 4 noodle pieces to 15½" for the front, back, left, and right arms, and one piece to 5½" for the battery mount. The battery mount will be glued to the side arms and the side arms will be glued to the front arms, so the ends of these pieces need to be curved. You should be able to dry-fit the frame together without any large gaps.

2. REINFORCE THE ARMS
Make a ¼" deep slit along the top of each of the 4 arms. Insert the 12" carbon fiber rods into the slits on the left and right arms, and the 15" rods into the front and back arms. Squeeze glue into the slits so that there are no dry areas. This is important, as gaps will allow the arms to flex.

3. ATTACH THE ARMS
Measure and mark 5" from either end of the front and back arms. Cover the entire curved mating surfaces of the left and right arms with glue and center them on the marks. Do the same with the battery mount; it fits centered between the left and right arms. Run strapping tape along the bottom of the 4 arms.

4. MOUNT THE MOTORS
Attach the motors to plywood scraps and secure to the ends of the arms using strapping tape. (Zip ties don't work well here.)

5. MOUNT THE BATTERY AND ELECTRONICS
Cut enough foam out of the bottom of the battery mount to fit a 3S 2,200mAh battery. Secure battery with a velcro strap. Glue the electronics to a piece of scrap plywood or plastic and attach to the top of the battery mount with more velcro straps.

Configure the flight electronics as you would for any quadcopter (see our guide at makezine.com/the-handycopter) and take 'er out for a test flight. You can add lights if you like for night flying; LED strips fit perfectly into the hollow noodles. Improvise, have fun, and don't be afraid to try out new ideas!

Follow the complete step-by-step instructions and share your noodle drone at makezine.com/behold-the-noodle-copter

8. MOUNT THE RECEIVER AND FLIGHT CONTROLLER

Attach your flight controller to the upper body plate using hot glue, double-sided tape, or bolts through the mounting slots. (We used the Flip 1.5 MWC controller. You can download the settings at the Maker Hangar project page.)

Bind your R/C receiver to your transmitter (see Maker Hangar Season One, Episode 12), and then set the throttle ranges by plugging each of your ESCs, in turn, into the receiver's Throttle port (Season 2, Episode 4). Mount the receiver and plug it into the flight controller (Figure U). Center the yaw servo and tighten the linkages.

Finally, screw the top plate onto the standoffs to protect your electronics (Figures V and W), and your Maker Hangar Tricopter is complete! ⬥

ABOUT FLIGHT CONTROLLERS

The flight controller board converts the signals from your transmitter into the motor speeds that move your tricopter. It also reads the aircraft's position and movements with its onboard gyros and accelerometers, and makes tiny changes to motor speeds to counter the wind, torque, and other forces that are trying to tip the copter over.

These are the boards I recommend for the Maker Hangar Tricopter:

- **OPENPILOT CC3D** — the best flight experience, easy setup, but tuning takes time
- **HOBBYKING KK2** — OK flight experience, fast tuning with onboard display, best for beginners
- **ARDUPILOT APM 2.6** — most powerful and expensive; programmable waypoint capabilities with GPS, compass, and barometer
- **FLIP 1.5 MULTI WII CONTROLLER (MWC)** — small, simple, and affordable, but powerful and flies well; optional barometer and compass

To learn more about flight controllers and how to fly your tricopter, watch the complete Maker Hangar how-to video series at makezine.com/go/makerhangar!

SPECIAL SECTION

DRONES | BUILD YOUR FIRST TRICOPTER

isolated from vibrations by short wire ropes. Clamp the 4 wire ropes into the brackets on the bottom plate, but don't connect the camera tray yet (Figure N).

5. INSTALL THE ESCS

Connect the 3 electronic speed controllers (ESCs) to the motors and zip-tie them to the arms (Figure O).

Arrange the 3 arms in their folded configuration, then measure out enough wire to extend all the power and ground wires to meet at the back of the body (Figure P). Solder the extension wires and insulate connections with heat-shrink tubing. Strip the free ends and solder them into your battery connector (Figure Q).

NOTE: Now's also the time to splice in a JST connector (optional) if you want to power an onboard FPV (first-person video) system and watch live video from the tricopter. Learn more about batteries, FPV, and other flight components in the first season of Maker Hangar videos.

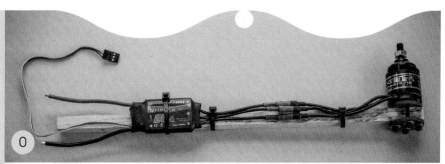
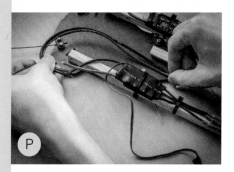
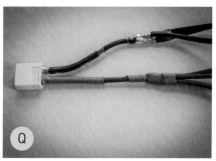

6. ATTACH THE ARMS

Bolt the 2 front arms to the lower body plate through the outer mounting holes, using M3×25mm bolts and lock nuts. Place the upper body plate on top, then pass 2 more bolts through the locking slots and the inner arm holes, and secure with washers and lock nuts. Finally, clamp the tail arm between the body plates using 4 bolts (Figure R).

Test the folding action and loosen or tighten bolts until the arms fold smoothly and lock forward securely.

6. MOUNT THE LANDING GEAR

Zip-tie the 2 plywood landing struts to the front arms (Figure S).

7. SUSPEND THE CAMERA TRAY

Clamp the free ends of the wire ropes into the brackets on the camera tray. Make sure the camera platform faces forward (Figure T) and the bolt heads face outward; you'll need access to them to adjust the tray later. Strap the battery onto the tray with the velcro strap.

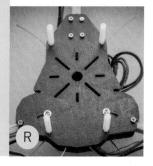
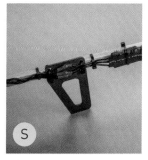
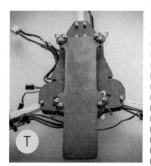

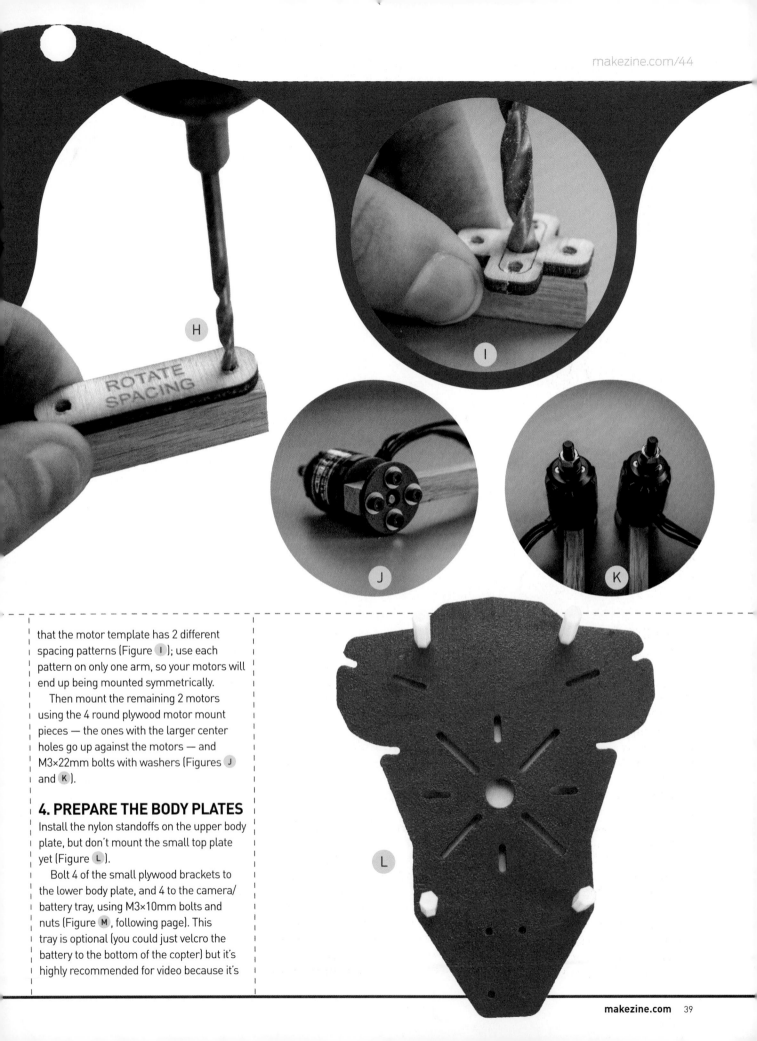

that the motor template has 2 different spacing patterns (Figure I); use each pattern on only one arm, so your motors will end up being mounted symmetrically.

Then mount the remaining 2 motors using the 4 round plywood motor mount pieces — the ones with the larger center holes go up against the motors — and M3×22mm bolts with washers (Figures J and K).

4. PREPARE THE BODY PLATES

Install the nylon standoffs on the upper body plate, but don't mount the small top plate yet (Figure L).

Bolt 4 of the small plywood brackets to the lower body plate, and 4 to the camera/battery tray, using M3×10mm bolts and nuts (Figure M, following page). This tray is optional (you could just velcro the battery to the bottom of the copter) but it's highly recommended for video because it's

SPECIAL SECTION

DRONES | BUILD YOUR FIRST TRICOPTER

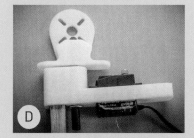
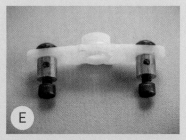
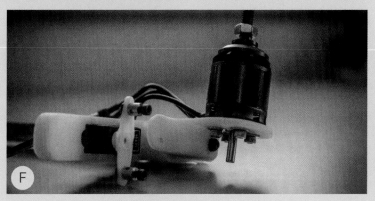
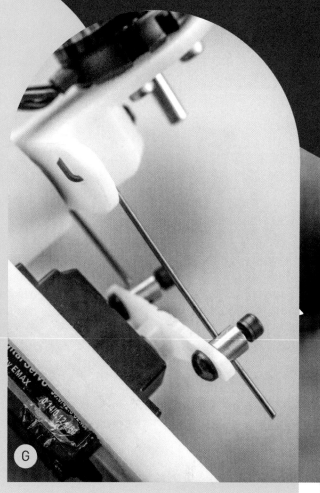

A TRICOPTER FOR MAKERS

The Maker Hangar Tricopter is made of wood — hackable, easy to drill and cut, and a natural absorber of vibration, the enemy of aerial video. The airframe is big, with plenty of room for large controller boards, video transmitters, drop mechanisms, or whatever you can imagine. And we widened the front arms to about 150° so our tricopter is more agile.

The kit includes a 3D-printed tail assembly and all the hardware you'll need, plus a wire rope vibration absorber that will pretty much erase camera vibrations even if your propellers are unbalanced. A carbon-fiber hinge provides a strong, smooth connection between the tail motor and airframe.

Finally, like most tricopters, the two front arms lock in place for flight, then fold back neatly for transportation and storage.

It's a great kit for anyone wanting to get into multicopters or aerial photography. You can also build it totally from scratch: download the PDF plans, laser cutter layouts, 3D files for printing, flight controller settings, and watch the how-to video series at makezine.com/go/makerhangar.

1. SAND AND PAINT

Sand down any burrs or splinters on the wooden parts. If you wish, paint with a couple of light coats (Figure B).

2. ASSEMBLE THE HINGED TAIL

To build the hinge, glue the 2½" carbon rod flush into the ¾" carbon tube using CA glue (Figure C). Hot-glue this end into the 3D-printed motor mount. Also hot-glue the 1" carbon tube into the 3D-printed tail piece.

Now put it together: Slide onto the hinge rod an M4 washer, then the tail piece, then another washer. Finally, glue the ½" carbon tube to the end of the rod to capture the whole assembly.

Hot-glue the servo into the tail piece (Figure D) and install 2 "easy connectors" in 1/16" holes on the servo arm (Figure E). You can glue the hardwood tail arm into the tail piece now as well.

Bolt the tail motor into the motor mount with M3 washers (Figure F).

Finally, connect the servo linkages. Use pliers to create a tiny "Z-bend" on the end of each push rod. Hook the bent ends into the motor mount, and slide the unbent ends into the easy connectors on the servo arm (Figure G).

3. ASSEMBLE THE FRONT ARMS

Drill each front arm using the 2 templates provided: at one end for the motor mounts, and at the other end for the rotation bolts for folding the copter arms (Figure H). Note

makezine.com/44

Time Required: A Weekend
Cost: $300–$400

Materials

Maker Hangar Tricopter Kit $85 from lucasweakley.com/product/maker-hangar-tricopter-kit, includes:
- Laser-cut plywood airframe parts
- 3D-printed tail assembly
- Carbon fiber hinge pieces
- Oak square dowels, 7/16"× 7/16"×12" (3) for the arms
- Bolts, stainless steel, M3: 25mm (8), 6mm (4), 10mm (16), and 22mm (8)
- Lock nuts, M3 (25)
- Washers: M3 (16) and M4 (2)
- Bolts, nylon, 6-32×3/8" (4)
- Nuts, nylon 6-32 (4)
- Standoffs, 6-32×1½" (4)
- Cable ties (20)
- Push rods, 2½"×0.047" (2)
- Push rod connectors (2)
- Velcro straps (2)
- Wire rope, 3" lengths (4)

Electronics (not included) — see the kit web page for complete recommendations:
- Flight controller board see page 41
- R/C receiver to match your R/C transmitter
- Motors, brushless outrunner, 900kV (3) Emax GT2215/12
- ESCs, 20A (3) Emax Simon
- Props, 10×4.7 (3)
- Batteries, LiPo, 3,300mAh (2)
- Servo, micro
- Servo extension, 6"
- Wire, 16 gauge stranded
- Heat-shrink tubing
- Servo cable, male to male
- JST connector (optional)

Tools
- Drill and bits
- Pliers, needlenose
- Pliers, side cutting
- Wire cutters/strippers
- Hot glue gun
- Cyanoacrylate (CA) glue
- Screwdriver
- Hex driver set
- Adjustable wrench
- Sandpaper
- File
- Hobby knife
- Soldering iron and solder
- Heat gun or hair dryer
- Helping hands (optional)

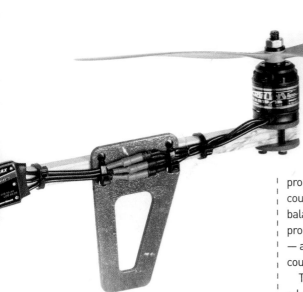

QUADCOPTERS ARE A LITTLE EASIER TO BUILD, BUT TRICOPTERS HAVE ADVANTAGES that make them more exciting to fly — especially for shooting aerial video. I built my first one in 2010, inspired by David Windestal's beautiful aerial GoPro videos (rcexplorer.se/fpv-videos-setups). I didn't get many flights out of that first build, but I learned a lot. After building several more, I've developed an affordable kit that anyone can build — the Maker Hangar Tricopter.

WHY FLY TRI?

A tricopter's three motors are usually separated by 120°, not 90° like a quadcopter's. This makes them great for video because you can place the camera really close to the body and still have no propellers in view. And where quads must rely on counter-rotating propellers to handle torque and balance the aircraft, a tricopter can use identical props because it has a special servo in the back — a yaw servo — that twists the tail motor to counter torque (Figure A).

Tricopters fly differently too. With their dedicated motor for yaw (turning), they fly with more fluid, natural-looking movements — they can bank, pitch, and yaw like an airplane, but still hover like a helicopter. A quadcopter's flight is more robotic, as the controller board calculates the precise rotation for all four motors to create the proper torque and balance to yaw the aircraft. If you let go of the stick, a quad stops turning abruptly; for video work, this can be obvious and distracting. Let go of a tricopter's stick and the tilted tail motor takes a moment to return to a hovering position; this gives you a slow stop and even a little overshoot, as though a person were moving the camera.

Finally, tricopters are a lot of fun to fly, especially for stunts and acrobatics. The tilting motor also gives you much higher yaw speeds — that means they turn faster.

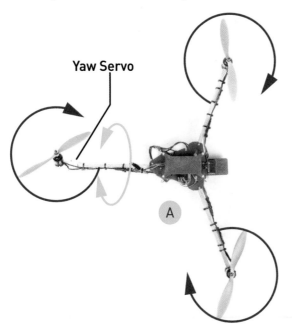

Yaw Servo

SPECIAL SECTION

DRONES | BUILD YOUR FIRST TRICOPTER

WRITTEN BY LUCAS WEAKLEY

BUILD YOUR FIRST TRICOPTER

LUCAS WEAKLEY is studying aeronautics engineering at Embry Riddle Aeronautical University. He also makes and sells aircraft kits at lucasweakley.com. He's a certified AutoCAD draftsman, an Eagle Scout, and the host of *Make:*'s Maker Hangar video series at makezine.com/go/makerhangar.

THEY FLY SMOOTHER AND MAKE BETTER VIDEOS THAN QUADS. BUILD THE MAKER HANGAR TRICOPTER AND SEE FOR YOURSELF!

SPECS
- Flight time: 12 minutes
- Frame weight: 325g
- Flight weight: 1kg
- Compatible with 8"–10" props
- Wire rope vibration absorber
- 22mm motor mounts

36 makershed.com

raise the throttle until the motor starts to spin. You will be able to see which direction the motor is spinning and can now lower the throttle. Consult your flight controller documentation to verify the motor is spinning in the correct direction. If it's not, unplug the battery and swap 2 of the connections between the motor and ESC to reverse the motor rotation.

E. Repeat for the remaining 3 motors.

6. FLIGHT CONTROLLER AND RADIO SETUP

Most flight controllers come preprogrammed for an X-quad layout. If yours does not, you will need to configure it over USB or onboard controls, depending on the flight controller. For beginner pilots, it is recommended that you verify that the default flight mode is attitude/stabilized/auto-level. The flight controller's configuration software will also allow you to confirm that all the radio channel inputs are in the correct order or if any need to be reversed.

7. TEST FLIGHT AND TROUBLESHOOTING

Charge your battery and attach the propellers to the correct motors. Your first test flight should be used to verify that all components have been properly configured. Slowly raise the throttle until the quad almost leaves the ground. If it shakes while hovering you will need to reduce some gains in the PID tuning.

If the quad lifts off, verify the radio channels are correct. This can be performed while the quad is still on the ground and the motors are spinning. For a mode 2 radio:

A. Move the right stick up (elevator), the quad should pitch forward.
B. Move the right stick left (aileron), the quad should pitch to the left.
C. Move the left stick to the left (rudder), the quad should rotate counterclockwise.

TAKE FLIGHT

Your quad is now ready for practice flights — refer to our Multirotor Checklist on page 29, and only fly in areas where you can safely control your rig without harm to others. ◉

Get detailed instructions at hovership.com/guides

SPECIAL SECTION

DRONES | HOVERSHIP

ANATOMY
OF A QUADCOPTER

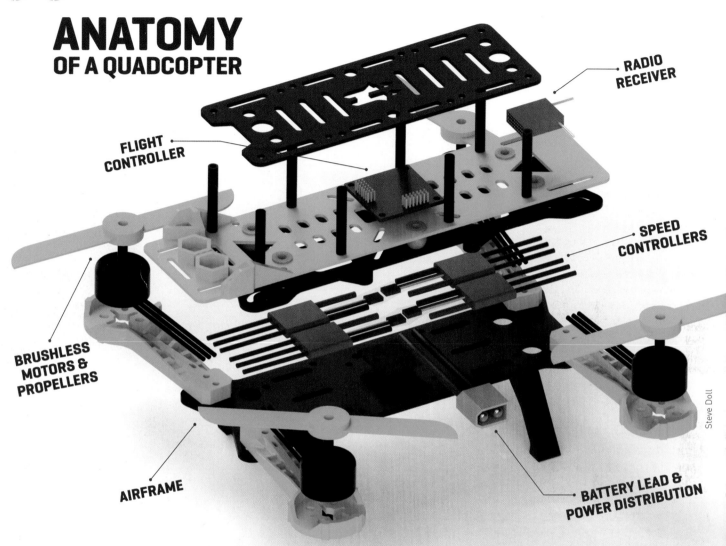

- FLIGHT CONTROLLER
- RADIO RECEIVER
- SPEED CONTROLLERS
- BRUSHLESS MOTORS & PROPELLERS
- AIRFRAME
- BATTERY LEAD & POWER DISTRIBUTION

Steve Doll

motor or ESC wires to length, leaving some slack, and use a wire stripper to expose the wire inside. Solder the wire into the gold bullet connectors. Male connectors should be placed on the motors, female connectors on the ESCs. Cover each connector in 4mm heat-shrink tubing (Figure D).

3. FLIGHT CONTROLLER AND RADIO RECEIVER

If your frame has an additional plate between the ESCs and the flight controller, assemble this part now. Feed the servo cables through to the center of the frame where the flight controller will be mounted. Before mounting your flight controller, you will need to determine which ESC plugs into the correct motor output of your flight controller. Consult the flight controller documentation for the order of an X-quad layout.

Once the ESCs are connected, attach the flight controller to the frame (Figure E). If your frame has a specific side that is the front (look for an arrow), your flight controller's front should be pointed in the same direction.

Consult your flight controller's documentation for the wiring between the radio receiver and flight board.

4. FINAL FRAME ASSEMBLY

Secure any loose wires or components with cable ties. Assemble any remaining frame parts such as the top plate or camera mounts (Figure F).

5. MOTOR CALIBRATION

Consult your radio documentation for the procedure to bind your transmitter to the receiver. Once the radio is bound, plug one of the ESC servo connectors into the throttle channel of the receiver. With the propellers off, perform the following steps:

A. Turn the transmitter on, move the throttle stick all the way up.
B. Connect power from the battery to the quad. The ESC will make a series of beeps when powered on.
C. After the tones complete, lower the throttle. The ESC will make another series of beeps confirming the calibration.
D. Place a propeller on the motor you calibrated but DO NOT secure it. Slowly

READY TO JUMP INTO THE WORLD OF FIRST-PERSON-VIEW (FPV) DRONE RACING? Building your own mini quad is a good way to get your first rig while learning the fundamentals of multirotor flight and piloting, as well as brushing up on your electronics and soldering skills.

A quadcopter is the most common configuration and is widely supported by airframes tailored to racing, which require extra durability during crashes. This build is based on the Hovership (which can be purchased or downloaded and 3D printed from hovership.com), but is applicable to many FPV racers.

1. FRAME (PARTIAL) ASSEMBLY

Attach the arms of the frame to the center base plates. Mount the power distribution board to the center of the baseplate using double-sided foam tape or nylon screws. Place 2 speed controllers on each side of the power distribution board. Attach a motor to each arm (Figure A).

2. POWER SYSTEM

Solder the red and black 14AWG silicone wires to the primary power input pads of the power distribution board. Cut the primary power leads to about 130mm or longer. (The length will depend on where you attach the battery.) Cut 2 pieces of 6mm heat-shrink tubing and slide them down each power lead. Solder the power leads to their corresponding sides of the battery connector. Slide the tubing up to the battery connector and shrink them to cover the exposed solder connections (Figure B).

Solder the red (positive) wires from the electronic speed controller (ESC) to the corresponding pads of the power distribution board. Repeat for the black (negative) wires (Figure C).

If your motors and ESCs did not come with connectors presoldered, it is recommended that you do so; this will make it easier to replace a component in the future, if necessary. Trim the

⚡ CAUTION ⚡
If your frame is made from carbon fiber, which conducts electricity, any exposed connections touching the frame will cause shorts in the electronics.

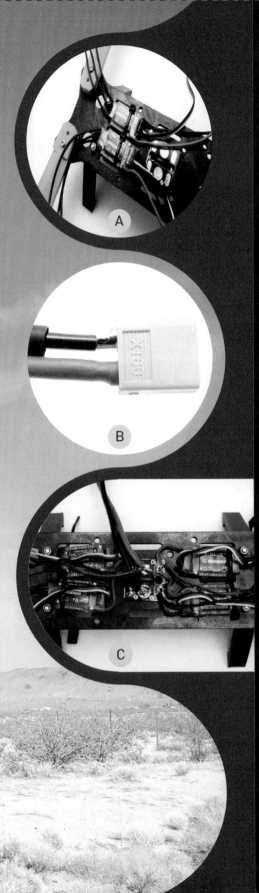

makezine.com/44

Time Required: A Weekend
Cost: $200–$400

STEVE DOLL is the founder and lead designer of Hovership. He has been building and flying multirotor helicopters for 4-plus years and enjoys creating aerial photos and videos.

Materials

» **Quad frame, mini, 250-300mm class**
» **Motor, brushless 1806-2204 size, 1900-2300kV** (4)
» **Electronic speed controller (ESC), 12A** (4)
» **Power distribution board**
» **Wire, silicone, 14AWG red and black**
» **Battery connector, male**
» **Heat-shrink tubing, 6mm and 4mm sizes**
» **Connectors, gold bullet, 2mm gold** if motors and speed controller do not include them
» **Propellers, 5×3 or 5×4** (2 clockwise rotation, 2 counterclockwise) spares recommended
» **Multirotor flight controller,** such as Naze32, OpenPilot, Multiwii, Pixhawk, KK2.1
» **Aircraft radio system, 4+ channel** 2.4GHz recommended
» **Lithium polymer battery, 3-cell, 1300-1800mAh capacity** more than one recommended
» **Battery charger.** LiPo
» **Velcro strap** for securing battery
» **Cable ties, small 4"-6" length**
» **First Person Video kit, 5.8GHz (optional)**

Tools

» **Soldering iron**
» **Heat gun or other heat source** for heat tubing
» **Hex wrench set, metric**
» **Pliers or small adjustable wrench**
» **Wire strippers**
» **Double-sided foam tape or nylon screws**

SPECIAL SECTION

DRONES | HOVERSHIP

HOVERSHIP
3D-PRINTED RACING DRONE
THERE'S NO BETTER WAY TO START RACING THAN TO BUILD YOUR OWN FPV QUADCOPTER

WRITTEN BY STEVE DOLL

RULES

Multicopters must:
- Meet the criteria for their class
- Be piloted exclusively via FPV
- Be capable of taking off and landing vertically

Pilots must not:
- Interfere with any other pilot or his / her equipment
- Walk onto the course while a race is in progress
- Fly multicopters close to other pilots or spectators
- Be intoxicated

COURSES

Find a large, open space to set up your course. If flying on private property, you will need permission from the owner.

Forests and parking garages offer more challenging and exciting races for experienced pilots. However keep in mind that solid objects such as trees and pillars may block video signals.

You can create pylons or arches using pool noodles stuck into the ground with poles or PVC pipes. To keep races interesting, try to incorporate different types of turns — such as hairpins, slaloms, sweepers, and chicanes — straightaways, and various over-and-under obstacles. Some people lay Coroplast arrows on the ground as course markers.

Typically a race will consist of at least three laps around the course, but this is at the discretion of the organizers. Most races last between two and four minutes.

SAFETY

- Choose a secluded location where people won't inadvertently walk onto the course
- Handle LiPo batteries carefully and dispose of damaged cells
- We recommend that children not be present at racing events

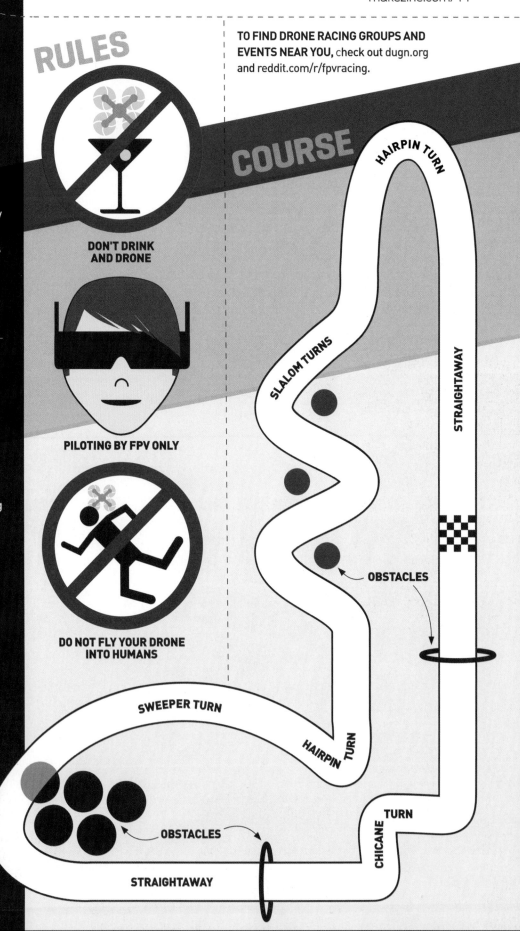

TO FIND DRONE RACING GROUPS AND EVENTS NEAR YOU, check out dugn.org and reddit.com/r/fpvracing.

SPECIAL SECTION

DRONES | BUILD YOUR OWN FPV RACE COURSE

DRONE DERBY

Set up your own FPV race with these handy guidelines

WRITTEN BY DONALD HILLS

CLASSES

MICRO/150   MINI/250   OPEN

DONALD HILLS is the founder of Drone Hire, an international directory of commercial drone operators, and the Ultimate FPV Racing League, fpvracing.tv. He loves aerospace, robotics, and innovation.

FPV RACING IS AN EMERGING SPORT combining multicopters, live video, and high-speed racing. As of yet, there aren't any official rules, but here are some guidelines based on our experiences.

CLASSES

Most racers fit into one of three classes based on their size. Race organizers may implement additional criteria, such as weight limits, for safety reasons.

Micro / 150
Suited to beginners and indoor racing events.
- Up to 150mm (measured diagonally from motor to motor)
- 4 motors
- 2-cell LiPo battery

Mini / 250
Currently the most popular class for FPV racing.
- Up to 250mm
- 5" props
- 4 motors; 1806 or 2204 brushless
- 3- or 4-cell LiPo battery

Open
Fewer restrictions for faster races (and more spectacular crashes).
- Up to 300mm
- 6" props
- 4 motors

FREQUENCIES

Prior to the event, the organizers should provide participants with a list of available video channels. Pilots should then choose a unique channel and this should be noted on the list. On race day, pilots should switch on their goggles and ensure that nobody else is using the same channel before switching on their video transmitter.

DRONES | PILOT'S CHECKLIST

makezine.com/44

THE MULTIROTOR CHECKLIST

WRITTEN BY AUSTIN FUREY AND THE FLITE TEST COMMUNITY

The only thing worse than getting to the flying field and discovering you've forgotten something critical is getting your multirotor in the air and realizing you've overlooked an important detail like securing your props. We've compiled this checklist to help you remember all the little things that will keep you safely flying and having fun.

BEFORE YOU HEAD OUT

- [ ] Check that onboard and transmitter batteries are charged. *Optional*: pack charger
- [ ] Grab the correct transmitter. And bring the bind plug for your receiver, just in case
- [ ] Check the weather — especially wind speed and precipitation — at your local flying site

PRE-FLIGHT

- [ ] Position your antennas and range-check your aircraft
- [ ] Check that wires are secure on the receiver and control board
- [ ] Check props for signs of stress
- [ ] Tighten prop nuts
- [ ] Secure battery strap

FPV

- [ ] Check that camera batteries are fully charged and the SD card is empty
- [ ] Test FPV screen and/or goggles
- [ ] Check goggle battery charge
- [ ] If flying near others, call out the video channel you are using

CRASH KIT

It's important to have a safe area, free of people, to make mistakes and, in the event of a crash, to be able to make simple repairs out on the field. Everyone's crash kit is going to be a little different, but here are some recommended items to bring with you every time you fly. It's easiest to incorporate them into a tackle box you can grab so everything is ready to go when you walk out the door.

- [ ] Extra propellers and prop nuts (plus the proper sized wrench or screwdriver)
- [ ] Clear tape, electrical tape, and hot glue sticks and gun
- [ ] Soldering iron, battery- or butane-powered if you don't have access to power
- [ ] Zip ties of multiple sizes
- [ ] Rubber bands
- [ ] Spare booms and landing gear
- [ ] Fire extinguisher and first-aid kit

Make sure you fly only in places where such activities are permitted. Get permission from property owners and avoid no-fly zones. In the U.S., these include:

- 5 mile radius around major airports
- 1 mile radius around large stadiums
- National parks
- Military bases

Other local restrictions may apply. For an interactive map, see mapbox.com/drone/no-fly

SPECIAL SECTION

DRONES | FORMULA FPV

Around 80 flyers turned out in mid-January for an FPV racing and combat games Meetup hosted by Game of Drones and the Aerial Sports League at a port-side park in Oakland, California.

downloading software, matching rotors, and dealing with technical difficulties, it took around four hours to set up, he says. "There's a tremendous amount of time to get up to speed with the products out there, how they work, dealing with the concept of a flight controller, PIDs [proportional, interval, and derivative values], and the receiver binding. It's a nightmare."

Kember's and Koblasa's company, West Coast Rotors, plans to offer customization and personalization of four popular models, as well as make them race ready. As of January, they had a backlog of orders, though they're approaching the company as a side project.

One of the reasons out-of-the-box racing drones aren't ready — and why such a service could be very popular — is that there is still so much experimenting going on. At every event I attended, racers brought cases full of components, spare parts, and extra batteries and spent 90% of their time getting ready and 10% racing. "We're still finding that sweet spot between the frame parts and the motor and the FPV gear," says Cornblatt. "There really isn't a clear winner yet, everyone's kind of still figuring it out."

"Half of this is building them, half is tuning them," concurs Kurt Somerville, who flies a mini deadcat-style quadcopter (pictured on page 28). Its frame, which he bought from a maker in China, is sort of a hybrid, with the front rotors situated straight across the frame, and the back splayed out behind, which he says is good for stability and hard banking, though it sacrifices some power due to smaller rotors.

"The cool thing right now is [FPV racing] is super underground," Somerville says. "You have these super close-knit groups that meet up to fly whenever they can." But it's only a matter of time, he adds, until we start to see organized — perhaps even televised — leagues.

Several groups are working toward making that a reality. Pellarin is recruiting pilots from France, Holland, and the U.S. (including Puertolas) to form a loose, informal league on an international scale. And in January, Doll was one of about 25 fliers who participated in a multirace event hosted by Aerial Grand Prix at the Apollo XI RC airfield in Los Angeles. An international organization with more than seven chapters, Aerial GP will be holding races throughout 2015 in the Netherlands, France, Australia, Mexico, and elsewhere, and hopes to further standardize the sport.

Puertolas also anticipates more organization in the near future. "I can see this being a big thing," he says. "I never saw somebody who tried it, tried the glasses on, and wasn't like, 'whoa!'"

SAFETY, SAYS PUERTOLAS, WILL BE CRUCIAL TO GROWING THE SPORT
For tips, see *"Multirotor Checklist,"* facing page, and *"Drone Derby,"* page 34.

batteries for the standard three, which enables greater speed and thrust. What's more, it means the drones can maneuver quicker, lifting over and accelerating around obstacles.

"We think [racing] has real meaning," Pellarin says. "We're sick of hearing about drones just because they're a toy of the NSA, or carrying bombs, or killing people. Drones can be fun; drones can be a mechanical sport for the next century. It requires skills, it requires engineering, it requires piloting."

Such competition can also lead to real advances in design, mechanics, and technique. Leagues and contests offer a sort of Darwinian incentive to building better drones, and require competing against — or collaborating with — people with different ideas.

Eli D'Elia and Marque Cornblatt, better known as the founders of the drone fighting ring Game of Drones, have gotten the bug as well. "FPV racing is a real visceral experience. It basically makes anybody who can do it kind of like Superman for 5 to 10 minutes," says D'Elia, who has a long history in action sports and in video games. To him (and many others), drones are bringing these worlds together. "In my mind, it has completely replaced video games, and has become my new addiction," he says. "If I could fly all day, first person video, I would."

When D'Elia and Cornblatt first started battling, there was an immediate need to rethink the drones so they wouldn't get quite so devastated upon collision. "It was kind of like the crucible of drone combat sports that birthed Game of Drones," says Cornblatt. "You saw every kind of engineering solution you could imagine, really high tech, really low tech. The problem was, no matter what, they all fell apart."

They tried frames of laser-cut cardboard, waterjet-cut carbon fiber, and Home Depot hardware; and they eventually realized they needed something truly durable: a military-grade polymer alloy body, which they now sell. "It's through the fighting, it's through the racing, and it's through going out every weekend and experiencing the challenges for the consumer that we've come up with a list of objectives that we really plan on solving," Cornblatt continues. He figures they can simplify drone racers down to a low profile, slick, crash-resistant sliver.

If they do, they'll be competing against Hovership's 3D-printed, 250-class racing drone called the MHQ2, and a newer laser-cut, carbon-fiber version called the ZUUL. It's only been in the past 12 months, creator Steve Doll says, that small drones — 250 millimeter mainly, though the 150 class is growing in popularity — became powerful enough to support the FPV gear needed to race.

Doll, who's been building and flying drones for years, gave the MHQ an H-shaped body, rather than the more typical "x" or "+" shape, so the keychain-size security camera, linked to a 5.8GHz band video transmitter, can sit on front, out of the way of the 5-inch rotors. Lightweight with no lag, keychain cams are the preferred choice. They also have a good light sensor, meaning they respond quickly to changes such as going from shade to sun.

Doll doesn't see a lot of carryover from racing to industry, which has been focused of late on GPS guidance, obstacle avoidance, and artificial intelligence, rather than piloting and first-person. "It's really just a hobbyist thing, taking this thing and having fun with it," he says. "All these motors and speed controllers were designed for R/C airplanes. It's all still very much a hack, in a way. We're kind of just making this up as we go along."

The same is true of technique and rules. "The best rule," says Pellarin of Airgonay, "is no rules." Still, while it may not be explicitly prohibited, racers tend to avoid contact (unlike drone combat), as any collision with an opponent is likely to knock both drones out of the race.

As this all gets dialed in, Elliott Kember and fellow racer Tyler Koblasa see a niche for a tuning and support company. Koblasa races a hexacopter called the TBS Gemini, which was designed by Team Blacksheep to be ready out-of-the-box. But between mounting hardware, preparing flight controllers, receivers, and video,

> "FPV RACING IS A REAL VISCERAL EXPERIENCE ... IF I COULD FLY ALL DAY, I WOULD"

SPECIAL SECTION

DRONES | FORMULA FPV

Almost any open space can become a drone racing arena. Organizers set up hoops, flags, and poles to create challenging courses, or use existing obstacles such as buildings or trees.

body rotating to face the direction they want to go while the drone is still sliding sideways through the air. The racers sit on the ground or in folding chairs, concentrating only on the controllers in their hands and the horizon.

"We look like Ray Charles playing the piano," remarks Kember.

It's a race to four laps, but really, it's a race of submission. One drone crashes into the ground, tumbling like a downhill skier; another gets lost and flies off course. (FPV has virtually no peripheral vision so if a pilot loses sight of the route markers it's very difficult to find them again.) The two remaining contenders fly around the course a few times before one — Kember's — laps the other and is declared the winner by consensus.

FPV racing is the hot new thing in competitive drone flying. Over the past year or so, more and more pilots have started building racers. The craft are typically small, compact quadcopters, measuring less than 10 inches diagonally from rotor to rotor.

It's taking off partly because it's fast, fun, and accessible. People can imagine doing it themselves, says Carlos Puertolas, who is described by many as the best drone racer in the Bay Area, if not the world. The technology is getting cheaper and easier; a pilot now can get his or her start for a few hundred dollars, and out-of-the-box racers are starting to become available.

Puertolas, who is sponsored by Luminier, flies their frames, produces video, and consults on design for the racing quads the company makes. But still he recommends building your own: "There's nothing wrong with getting one that is already built, but you get a lot of knowledge from building it yourself, not only to be able to repair it but to be able to troubleshoot the problems

> "YOU DON'T NEED MUCH SPACE TO ORGANIZE SOMETHING AMAZING FOR THE PILOTS."

that you have, and understand what's wrong."

Last fall, European enthusiasts calling themselves Airgonay became internet famous for videos featuring FPV drones ripping around a course, set up in a French forest, at nearly 70 mph. Like FPV Explorers in the U.S., the group is made up of people from varied backgrounds and professions who get together whenever they can to race informally.

"This is actually our lunch break. This is what we do every day," says Airgonay founder Hervé Pellarin, who does marketing for FPV companies Fat Shark and ImmersionRC. "You don't need much space to organize something very amazing for the pilots, as an experience. Just a piece of forest, trees, and path, and you can have fun all day."

Like many racers, Pellarin overpowers his drones, substituting larger motors and rotors and higher voltage four-cell

FORMULA FPV.

DRONE RACES ARE TAKING OFF

WRITTEN BY NATHAN HURST

LITTLE PROPS BUZZ AT DIFFERENT PITCHES, LIFTING DRONES INTO THE CHILLY AIR OVER CÉSAR CHÁVEZ PARK ON A GRAY, BREEZY DAY IN BERKELEY, CALIFORNIA. A dozen pilots prep equipment on the open grassy space that juts into the San Francisco Bay, replacing prop blades or setting up their first-person-video (FPV) rigs.

"Who just turned on?" calls out Elliott Kember, who's getting video interference. "Someone's on my channel."

Nearby, a series of PVC poles wrapped in foam pool noodles demarcate a makeshift racecourse a couple hundred feet in length. There are several hairpin switchbacks, two 90-degree turns, and a low arch, all marked in bright orange so the pilots can see them through cameras mounted on their specialized drones.

The group, a Meetup called FPV Explorers & Racers, formed last summer. They get together about once a week, usually on Sundays, to hang out and fly drones.

With their gear finally ready, four racers set their drones on starting mats beneath the arch. Each dons goggles that show real-time video from keychain cameras mounted on their crafts. Somebody counts down from three and the drones lift off and rip around the first turn.

They're fast — most can exceed 50 mph — but what's more impressive is their acceleration. They corner like a drift car;

Hep Svadja

SPECIAL SECTION

DRONES | FORMULA FPV

makezine.com/44

Flying your quad in your backyard is great fun. But flying with friends — whether to share displays of acrobatics, engage in friendly races, or conduct first-to-crash jousts — makes the endeavor infinitely more enjoyable. Groups of fliers around the world are gathering now to show off their craftsmanship and aerial prowess. Leagues are being formed, rules are being set, and superstars are being made. The best part? There's room for more. Get your rig and join in. It's a multirotor revolution. ◐

SPECIAL SECTION

DRONES | ROTARY CLUB

DON'T DRONE ALONE
ROTARY CLUB

Maker Pro Q&A

TINYCIRCUITS IS AN OPEN-SOURCE HARDWARE COMPANY SPECIALIZING IN DESIGNING AND MANUFACTURING VERY SMALL ELECTRONICS. Founder and president Ken Burns started the company as a side project in a spare bedroom of his house in 2011. Now housed in an old tire factory in Akron, Ohio, it sells more than 40 tiny products and employs seven people.

You once wrote an article for *Make:* about using your boards to track your outdoor cat, Conley. But you didn't say what you learned about his whereabouts. What do you now know about Conley?
People liked that idea. I estimate that there are now 200 to 300 cats that have been tracked with our boards. I learned that he's lazy. That he spends most of the time sleeping in a neighbor's shed down the road.

What changes when boards get tiny?
Our stuff is based on the Arduino, and it works like a regular Arduino. But by making it small, it allows you to put it in places you've never been able to before, like in very small rockets. You can power and control little Matchbox cars and prosthetic hands.

Are any of your customers spies?
One guy who works here used our technology to make a little spy device that he snuck onto my bookshelf. He used it to turn my air conditioner on and off, randomly, very quickly. That was pretty annoying, until I figured it out.

You manufacture TinyCircuits in the U.S. How do you make that work?
For us, manufacturing here in Akron costs less than outsourcing to China. In San Francisco we wouldn't be able to do that. Just the cost of labor and space would make that impossible. A software startup makes more sense in San Francisco, but for a hardware startup, the Midwest is great. Also, it's really not that hard to get qualified engineering people here and pay them a livable wage. We have some great engineering colleges nearby: Carnegie Mellon, Case Western, Akron University.

What's the biggest challenge when you're marketing tiny technology?
The biggest challenge for all hardware projects happens once the Kickstarter period is over. What typically happens is that the first month after you fulfill your Kickstarter orders, you do about one twentieth of what you did before. That's tough. So if you've had a $110,000 Kickstarter, like we did, then in the month after that we made about $5,000 in revenue. Pretty sharp drop off.

It's been called the "Valley of Death."
Because you have that initial money, and you know how many units you have to build. So you do that. But then you have to figure out how you're going to finance the next round. It's tricky. We initially raised $110,000, and we turned that into $50,000 worth of debt by the time we actually delivered the product. I don't know anybody who's actually made money on Kickstarter.

How did you get out of the Valley of Death?
We went to a lot of Maker Faires and talked to people. We blogged a lot and posted on Facebook. If you can get independent people to buy your boards and write about them, it has tremendous cred. We've been building that over the last two years. It's slower, organic growth.

All your hardware is open source. How does that work for you?
It opens up a lot of advantages. Our hardware is not that complicated. If someone wanted to copy it, it wouldn't be difficult, so closing it off wouldn't give us that much protection. By opening it up, we're able to play in this much bigger community. And the cool thing is that it really feeds this innovation engine: People use it, add something, and contribute it back; and the pool gets bigger. That enables really rapid innovation.

> *I estimate that there are now 200 to 300 cats that have been tracked with our boards.*

DC DENISON is the editor of the *Maker Pro Newsletter*, which covers the intersection of makers and business, and the former technology editor of *The Boston Globe*.

For more Maker Pro content, visit makezine.com/category/maker-pro.
Subscribe to the Maker Pro Newsletter at makezine.com/maker-pro-newsletter.

FEATURES

Itty Bitty Boards
Written by DC Denison

Meet the founder and president of TinyCircuits

Tim Hollister Photography

makezine.com/hands-on-health-care

can be dislodged, and patients can lose their sense of smell, taste, or even sight. Thinking about her thyroid surgery, she and Balzer wondered if a similarly noninvasive procedure might be possible.

Balzer downloaded a free software program called InVesalius, developed by a research center in Brazil to convert MRI and CT scan data to 3D images. He used it to create a 3D volume rendering from Scott's DICOM images, which allowed him to look at the tumor from any angle. Then he uploaded the files to Sketchfab and shared them with neurosurgeons around the country in the hope of finding one who was willing to try a new type of procedure.

Perhaps unsurprisingly, he found the doctor he was looking for at UPMC, where Scott had her thyroid removed. A neurosurgeon there agreed to consider a minimally invasive operation in which he would access the tumor through Scott's left eyelid and remove it using a micro drill. Balzer had adapted the volume renderings for 3D printing and produced a few full-size models of the front section of Scott's skull on his MakerBot. To help the surgeon vet his micro drilling idea and plan the procedure, Balzer packed up one of the models and shipped it off to Pittsburgh.

Balzer had unknowingly pioneered what researchers at the new Medical Innovation Lab in Austin, Texas, predict will soon be the standard of care. Using 3D printing to help plan procedures and to explain diagnoses to patients "is going to become the new normal," says Dr. Michael Patton, CEO of the Lab, which launched in October 2014 with the goal of bringing new ideas for medical devices and technologies to market. Patton says its doors are open to creative thinkers like Balzer and points out that 3D printing can accelerate the process of product, tool, and device development in medicine. "What you can now do through 3D printing is like what you're able to do in the software world: Rapid iteration, fail fast, get something to market quickly," Patton says. "You can print the prototypes, and then you can print out model organs on which to test the products. You can potentially obviate the need for some animal studies, and you can do this proof of concept before extensive patient trials are conducted."

Trials, tests, and studies are a key point: One of the important roles of Medical Innovation Labs is to help guide inventions through the regulatory process. "It's extensive and it's burdensome," Patton says, and it's a reason many great ideas never make it off the back of cocktail napkins. But Patton doesn't anticipate any regulatory issues with using 3D-printed models for surgical planning and predicts that other advances involving simple scanning and printing will be brought to market with relative ease. "That is part of the new frontier with scanning and 3D printing, and we don't see the regulatory hurdles that you would see with implants," Patton says. He looks forward to being able to scan a broken bone at home and print out a breathable cast.

Closer at hand, but no less fantastic, is a handheld medical imaging device that will use ultrasound scanners to generate 3D images — no MRI necessary — and send them to a cloud service, where they can be accessed by doctors around the world. A startup called Butterfly Network recently received $100 million in funding to build the device and the cloud tool, which will recognize and automatically diagnose certain irregularities — such as a cleft palate in an unborn fetus — and learn over time. As more scans are uploaded, it will be able to automate more diagnoses.

Patton says he's even more excited to work with inventors and makers than experts within the medical field. "So many people are trained to keep their head down and focus on practicing medicine," he says, "and sometimes they don't think about why they do things a certain way, or how they could do them differently." Balzer is a prime example. 3D scanning and printing made high-tech health care accessible to him, and it also allowed him to influence progress in the medical establishment. This, as Patton says, is a radical new model for medical innovation.

Balzer has, in fact, been developing a product for medical use

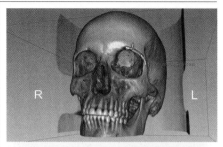
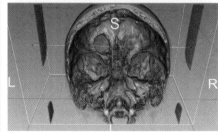

Lodged just behind Scott's left eye was a three-centimeter tumor

that's similar to the Butterfly Network device, combining portable 3D scanning with a platform for doctors and patients to share images via a secure (HIPAA-compliant) cloud server. He's also become more focused on education and hosts a podcast called *All Things 3D*, on which he often invites doctors to speak. Recently, he organized a free seminar on 3D in medicine. "My big message now is that this stuff is out there, and a lot of it is free," he says. "The first thing is getting the word out that your hands aren't tied. Your buddy's got a 3D printer? Use it."

Scott had the tumor removed at UPMC in May 2014 through a small opening above her left eye. The neurosurgeon discovered that the tumor was starting to entangle her optic nerves and told her that if she had waited six months, she would have had severe — and possibly permanent — degradation of her sight. The procedure took eight hours and 95% of the tumor was removed. She was back at work in three weeks. Her scars, Balzer says, are visible only to her. ●

PRINT YOUR OWN

Want to print *your* medical image? Ask your doctor for your DICOM files and download 3D Slicer (slicer.org). Then use the Region Growing tool to segment the image. Extract a 3D mesh of the surface, save as an STL, and use ParaView (paraview.org) to simplify it to a manageable number of triangles. To see more details, check out *Make:* Volume 42, *page 83*, or visit makezine.com/projects/3d-print-your-medical-scan.

FEATURES | Hands-on Health Care

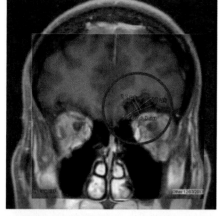

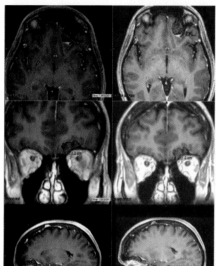

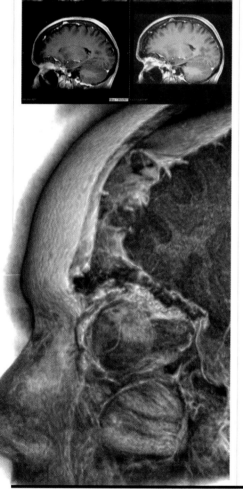

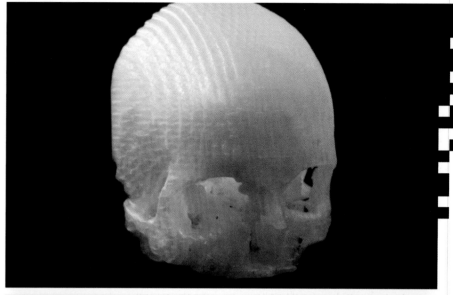

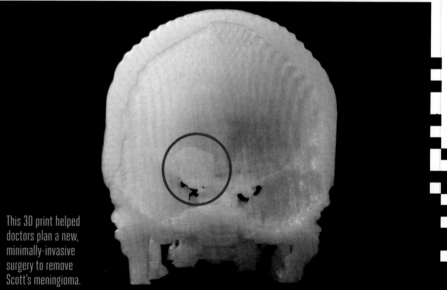

This 3D print helped doctors plan a new, minimally-invasive surgery to remove Scott's meningioma.

were heretofore unimaginable.

A team of British and Malaysian researchers used a multimaterial 3D printer to create model heads with realistically textured skin, skull bones, brain matter, and tumors so that students could safely practice high-risk surgeries. In the U.S., a pair of doctors at the University of Michigan printed customized tracheal splints for two young children with a condition called tracheobronchomalacia, or softening of the trachea and bronchi, which causes the airways to collapse. The splints will allow the tracheal muscles to develop, and as they do, the supports will be safely absorbed into the children's bodies.

But one of the most widely useful applications is one of the simplest: Using patients' CT scans to 3D print precise models of organs so that doctors can plan and prepare for surgical procedures. The software and equipment necessary are easily accessible to anyone — one University of Iowa surgeon tracked down a local jewelry maker with a 3D printer and convinced him to fabricate custom model hearts for the university in his spare time.

Balzer wanted a tangible model of Scott's cranium so that he could get perspective on the location and size of the tumor and think about what kind of treatment to pursue. The standard removal process for a tumor like Scott's, known as a meningioma, is a craniotomy, in which the skull is sawed open. Her tumor was located under her brain, so to remove it, doctors would have to physically lift her brain out of the way. This is as risky as it sounds. Nerves

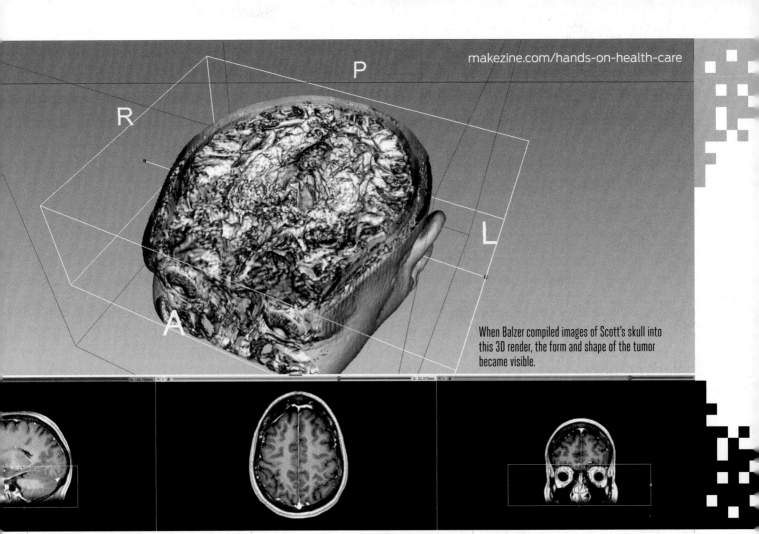

When Balzer compiled images of Scott's skull into this 3D render, the form and shape of the tumor became visible.

her left eye. They were understandably terrified, but neurologists who read the radiology report seemed unconcerned, explaining that such masses were common among women, and suggested Scott have it checked again in a year.

That didn't sit well with Balzer. Scott's recent thyroid surgery had taught them that getting the best care requires being proactive and extremely well informed. A typical thyroid removal is performed via a large incision across the throat that requires a long, uncomfortable recovery and leaves a big scar, but when he and Scott began looking for alternatives, they discovered that she could avoid all that if they traveled from their home in California to the Center for Robotic Head and Neck Surgery at the University of Pittsburgh Medical Center. There, surgeons perform delicate procedures with a robotic arm that scales down their movements, making them smaller and more precise than what the human hand is capable of alone. The experience familiarized Balzer and Scott with both the cutting edge of medical technology and the importance of doing

their own research. So although the first doctors told them to wait, Balzer and Scott sent the MRI results to a handful of neurologists around the country. Nearly all of them agreed that Scott needed surgery.

At this point, Balzer requested Scott's DICOM files (the standard digital format for medical imaging data) so he could work with them at home. It was a crucial step. A few months later, Scott had another MRI, and the radiologist came back with a horrifying report: The tumor had grown substantially, which indicated a far more grave condition than was initially diagnosed. But back at home, Balzer used Photoshop to layer the new DICOM files on top of the old images, and realized that the tumor hadn't grown at all — the radiologist had just measured from a different point on the image. Once his relief subsided, Balzer was furious and more determined than ever to stay in control of Scott's treatment. "I thought, 'why don't we take it to the next level?'" Balzer says. "Let's see what kind of tools are available so that I can take the DICOMs, which are 2D slices, and convert them into a 3D model." That decision

changed everything.

Balzer, a former Air Force technical instructor and software engineer as well as 3D-imaging aficionado, is probably better prepared than most to take medical diagnostic technology into his own hands, but it's not necessary to have his level of expertise to use 3D imaging to better understand a diagnosis and possible treatments, and it's only getting easier. Groundbreaking advances in medical care are being made using basic maker tools and software, which means that state-of-the-art health care is becoming cheaper, faster, and more widely accessible, but also — and perhaps more importantly — it means that we can use these same tools to make sure our own health care is up to par.

3D printing has already brought some astonishing changes to non-DIY medical care, and the field is still in its infancy. In China and Australia, where 3D-printed implants have been approved, doctors have replaced cancerous and malformed bones with bespoke titanium pelvises, shoulders, and ankles that are produced with speed, precision, and strength that

FEATURES

HANDS-ON HEALTH CARE...

SARA BRESELOR is a journalist and editor based in San Francisco. She writes about technology, art, culture, and weird news for the *Harper's* Weekly Review, *Wired*, and *Communication Arts*.

Written by Sara Breselor

WHEN HIS WIFE WAS MISDIAGNOSED, MICHAEL BALZER USED 3D PRINTING AND IMAGING TO GET HER WELL.

THE SUMMER OF 2013 FOUND MICHAEL BALZER IN GOOD HEALTH. A few years earlier, he'd struggled with a long illness that had cost him his job. As he recovered, he built an independent career creating 3D graphics and helping his wife, a psychotherapist named Pamela Shavaun Scott, develop treatments for video game addiction. Balzer's passion is technology, not medicine, but themes of malady and recovery have often surfaced during his digital pursuits. But Balzer didn't feel the full impact of that connection until that summer, shortly after he launched his own business in 3D design, scanning, and printing. In August 2013, just as the new venture was getting off the ground, Scott started getting headaches.

It might have been nothing, but Scott had gotten her thyroid removed a few months earlier, so the pair had been keeping an especially close eye on anything that might have indicated a complication. Balzer pestered his wife to get an MRI, and when she finally agreed, the scan revealed a mass inside her skull, a three-centimeter tumor lodged behind

All images courtesy of Michael Balzer

SWEET CITY BRENDANJAMISON.COM

The next time you take your tea with a lump or two, spare a thought for Brendan Jamison's sculptural work and the incredible potential of the humble sugar cube dissolving in your cup. According to Jamison, what initially began as "an attraction to the architectural properties of the brick-like form" grew into works that push the architectural limits of these commonplace materials.

Over the past decade, Jamison has constructed a myriad of elaborate structures, from a calf-high fireplace to a miniature Great Wall of China, using nothing but sugar cubes and adhesive. More recently he's partnered with sculptor Mark Revels to create even more ambitious works, including a series of cityscapes called Sugar Metropolis, comprised entirely of sugar-cube structures and installed in cities from New York and Los Angeles to Belfast.

"The beauty of working with sugar is that it is has so many associations and can operate on many different levels simultaneously," says Jamison. The shape, the texture, and even the imagined taste all feed into the viewer's experience.

— Andrew Salomone

MADE ON EARTH

CARDBOARD KESSEL RUNNER
THOMASRICHNER.BLOGSPOT.COM

Thomas Richner may have set out to clean his basement, but that stack of cardboard boxes was too appealing to send to the recycling bin. Instead, Richner spent more than 140 hours painstakingly shaping those cardboard boxes into an extremely detailed model of the Millennium Falcon from *Star Wars*.

Not only does it look fantastic, he matched the physical size of the model used during the filming of the iconic movies. He didn't skimp either; this cardboard model actually has retractable landing gear.

Richner, an animator by trade, really liked the idea of making something physical.

"With so many things going digital these days, I think another aspect that drove me to make this model was to create something tangible," he says. "There is a connectedness or appeal that happens when you can physically touch something that's difficult to achieve when it only exists as pixels."

— *Caleb Kraft*

REGAL RECEIVER KIPGEN.COM

Tom Kipgen crafts beautiful, old-school tube and crystal radio sets from rare parts and antique schematics. The woodworker and retired salesman has been at it for nearly two decades — building for collectors, friends, and family.

On his website you can see his unique designs: There's the Crystal Cyclops — a set that looks like a jukebox with a big circular broadcast display for the eye — and Canned Ham, described as "a tasty radio."

It's not just the designs that make the radios special — the B-1 Bomber crystal radio is made from the parts of a B-1 bomber. They're meticulously assembled from individual components — resistors, capacitors, diodes, and more — scrounged up from various electronics stores.

Constructing sets goes hand-in-hand with Kipgen's other wood-intensive craft: Building archtop guitars. Some radios take a couple of hours to make, others a couple of weeks.

"You don't see any unique looking radios anywhere," the 67-year-old says from his Oklahoma home. "I thought I could be creative with this."

— *Arvin Temkar*

MADE ON EARTH

The world of backyard technology

Know a project that would be perfect for Made on Earth? Email us: *editor@makezine.com*

GEAR UP: ROBOTICS

Your dream robotics project begins in the Maker Shed. From kits to books, we've got the bot for you.

> MAKERSHED.COM

Maker Shed
The official store of **Make:**

WELCOME

makezine.com/44

Is Game of Drones the Next X Games?

MARQUE CORNBLATT, THE CO-CREATOR OF GAME OF DRONES, KNOWS HOW TO GET YOUR ATTENTION. His contests, both indoor and outdoor, feature drone pilots going to battle. Cornblatt also creates popular videos, including an entertaining clip with a set of torture tests for a drone. First, he flew the drone into a glass windowpane, then let it free fall from 400 feet and crash to the ground. And most absurdly, the drone was the target for shotgun practice; his YouTube video, "Shotgun vs. Drone" has more than a million views.

Before he got into drones, Cornblatt had a video robot named Sparky that he brought to the first Bay Area Maker Faire. Each year, he improved Sparky, transforming it from an analog to a digital telepresence robot. "As a maker, I'd throw out last year's technology and start with new technology," he says. Sparky decreased from 300 pounds to about 6.

Cornblatt's craziest creation came next, the project he called WaterBoy and BucketHead. "For Burning Man, I wanted to come up with something that was as absurd as possible for the deep desert," he says. He wanted to seal himself in a suit filled with water, like the opposite of a diving bell. "I connected to people who had professional expertise in special effects and building wetsuits. I said: 'Here's what I'm trying to make.' They told me: 'No you can't do it. You're going to die.'"

He realized the suit — the WaterBoy half of the project — was a lot like a waterbed, and he found a Bay Area waterbed manufacturer who generously agreed to make it for him. Wearing the other half, BucketHead, he looked like a man who took an oversized goldfish bowl and stuck it on his head — with the water still in it. "I felt like I was a test pilot," he says. Cornblatt teamed up with the band OK Go at Maker Faire, and Damian Kulash went onstage in both WaterBoy and BucketHead, singing a song underwater, a truly remarkable performance.

Cornblatt came to drones through RC cars and planes. "I'm easily bored and I'm always looking for something to overcome it," He says. He started getting together with Justin Gray and other inventors in Oakland to "smash our toys together." Eventually, this became a weekly gathering they called Flight Club, and they began crashing drones on purpose. "I didn't want to fly drones by myself," says Cornblatt.

"The first thing we learned from setting out to crash drones was that commercial drones were super fragile and the parts were expensive," he says. He wanted to figure out how to make the airframe for drones more rugged. With co-founder Eli D'Elia, he launched a Kickstarter promising "to build an airframe that didn't need to be repaired." Cornblatt knew that he could produce good videos to help him raise the money and to gain momentum for what later became "Game of Drones."

During a contest, drones battle inside of cages and knock out their opponents, incapacitating them so that they cannot be quickly repaired by their pilots and returned to flight. Because some pilots were more interested in acrobatics than battles, he began adding competitions for pilots to demonstrate new tricks. Now Game of Drones includes racing where pilots wear FPV goggles. (See *"Formula FPV,"* page 24.) They are racing a new breed of drone — "tiny, fast, and angry like hornets."

Cornblatt, a race car enthusiast and an SCCA-trained driver, believes a competition like the Game of Drones is an important way to push the limits of a technology. Henry Ford was one of the early organizers of auto racing, staging match races and attempting to set land-speed records. His goal was the kind of publicity that would make automobiles popular — he wanted people to talk about what cars could do. When Ford started out racing, nobody thought of themselves as race car drivers. He had to recruit a competitive cyclist, Barney Oldfield, who was completely unfamiliar with the controls of a car. Nonetheless, by driving a one-mile track in one minute in a Ford vehicle, Oldfield was the first person to drive a car at 60 mph. Racing made Ford and his fortune.

Cornblatt wants people to get excited about drones, and he thinks that racing and other aerial games will do that. Events like Game of Drones are creating a new category of competition, much like the X Games did for skateboards and BMX bikes, developing a distinctive language of tricks and maneuvers, as well as promoting the broader interest in drones by more people — even among those who don't fly them.

BY DALE DOUGHERTY, founder and Executive Chairman of Maker Media.

DREAM. CREATR. EXCEED.

Leapfrog is known for bringing ideas to life fast. Now we bring them five times faster. Introducing the **CREATR HS**. With High Speed stamped into its name, the latest in our **CREATR** lineup is the largest, and yes fastest, 3D printer of its kind.

Backed by our best-in-class pricing, strength, versatility and incorruptible commitment to quality, the **CREATR HS** is the new standard for business, pleasure and everything in between. Are you ready to take your imagination full-speed ahead?

Creatr HS — 2015 SILVER iReviews

PRINTS BEST WITH MAXX

Leapfrog
3D Printers
lpfrg.com   Create The Future
US Sales Toll Free: 1.844.255.8455
Europe Sales: +31 (0) 172 503 625

Available from:

| coho3d.com | 3dprintlife.com | 3dbluemedia.com | 3dcadprinter.com | 3dprovensystems.com | 360tech.com |
| Seattle | Los Angeles | Atlanta | Atlanta | Albuquerque | Austin |

Distributed by: **wynit** We Distribute

E EPILOG LASER
ENGRAVE IT. CUT IT. MARK IT.
Laser cut even the most elaborate designs with ease.

Whether you're working with wood, acrylic, paperboard or just about anything else, Epilog Laser systems precisely cut the components you need.

Your vision combined with Epilog's power equals spectacular creations.

Assembly Required

Desktop Systems Starting at $7,995

Contact Epilog Laser today for a laser system demo!
epiloglaser.com/make • sales@epiloglaser.com • 888-437-4564

MADEiNUSA
Golden, Colorado

READER INPUT

A Master Maker in the Making

Allie with her Green Dollhouse, which earned her an "invention" award at her 3rd-grade science fair. Designed out of recyclable materials, it included green space, a rain barrel water collection system, and a space-saving collapsible wall bathroom. Allie also incorporated a working wind turbine and solar panel to run the light system, and added in some of her snap circuits to make a rechargeable battery storage.

» I am writing to tell you about my daughter. She LOVES *Make:* magazine. Her name is Allie and she is currently 9 years old. She hopes to become an engineer / inventor and work for NASA. She has had that aspiration since kindergarten.

She has never been a fan of dolls like other girls. She has been a Buzz Lightyear fan since she was 18 months old, when we had to break down and get her a Buzz doll because she constantly carried around an imaginary one.

It would be her DREAM to participate in a Maker Faire, or in some way be a part of *Make:* magazine. We were first introduced to *Make:* while visiting the Omaha Children's Museum. Allie caught the eye of some of the maker's working there and they gave her a copy of the magazine, and they said they wished she lived closer so they could work with her more. We can only visit once or twice a year since we do not live close by.

Thanks for encouraging and inspiring my daughter to keep making and inventing! —*Kara Weber, South Dakota*

To get Allie more involved, our Maker Faire Program Director, Sabrina Merlo, connected Kara and Allie with the organizers of both the Omaha and Des Moines Mini Maker Faires. Looking for a Maker Faire near you? Find them all at makerfaire.com

» **REALLY ENJOYED THE WEARABLES ISSUE OF** *MAKE:*

Just had to send you a note about the latest issue (Volume 43). Man, it had so many good projects! I bookmarked just the ones I wanted to do and there were like 10 stickies flapping around. [My wife] Jeanie liked the pieces on wearables designers as well. And the open-source RC airplane (*"Maker Trainer R/C Airplanes"*) from foam? That's unbelievable! Well, I'm sure the controller is still something beyond my meager skillz, but the plane seems doable.

But then the (*"Open-Source Smartwatch"*) project was pretty deceptively simple looking with the exploded diagram of only a few parts, but I went to the website and read the plans and browsed the github; boy, good luck to whoever tries cramming all that stuff together!

Anyway, I enjoyed the issue cover to cover. —*Brian Bruce, New York, NY*

» **OPEN SOURCE FOR BETTER 3D PRINTS**

The manufacturers of closed-source printers (MakerBot, Zortrax, etc.) should reconsider the benefits of open source for both the customer and the manufacturer. I will only buy open-source printers because they give me the flexibility to fix my own problems instead of being dependent on the manufacturer.

There are plenty of good open-source printers. Six out of 10 of the top performing printers in *Make:* magazine's annual guide (what *Make:* called "The Standouts") use open-source software. Four out of 10 have open-source hardware. If MakerBot Replicator 5th Generation and Zortrax M200 were open source, maybe a customer could figure out how to improve their test results. —*Ralph Kauffman, New York, NY*

Maker Profile: U of Nevada - Reno

Shown here with optional stand and accessories.

Students at the Engineering Robotics Lab at the University of Nevada-Reno use a Tormach PCNC 1100 to make prototypes and parts for fixed wing and quad rotor UAVs. "Using the Tormach CNC we are able to easily create complex wing shapes to develop wings that would otherwise be beyond our capabilities to make," explains Andy Smith, a graduate student in mechanical engineering.

To read the full story, go to

www.tormach.com/renodrones

UP mini $599

UP PLUS 2 $1,299

UP BOX $1,899

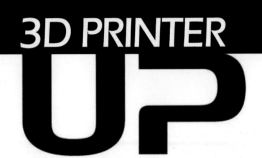

3D PRINTER UP

UP for Everyone

www.UP3D.com

Make:

> "We look forward to the day when drones are not a novelty, they're just considered a tool." — Eli D'Elia, co-founder, Game of Drones

EXECUTIVE CHAIRMAN
Dale Dougherty
dale@makermedia.com

CEO
Gregg Brockway
gregg@makermedia.com

CFO
Todd Sotkiewicz
todd@makermedia.com

CRO
Ed Delfs
ed@makermedia.com

CREATIVE DIRECTOR
Jason Babler
jbabler@makezine.com

EDITORIAL

EXECUTIVE EDITOR
Mike Senese
mike@makermedia.com

MANAGING EDITOR
Cindy Lum

COMMUNITY EDITOR
Caleb Kraft
caleb@makermedia.com

PROJECTS EDITOR
Keith Hammond
khammond@makermedia.com

SENIOR EDITOR
Greta Lorge

TECHNICAL EDITOR
David Scheltema

EDITOR
Nathan Hurst

EDITORIAL ASSISTANT
Craig Couden

COPY EDITOR
Laurie Barton

PUBLISHER, BOOKS
Brian Jepson

EDITOR, BOOKS
Patrick DiJusto

EDITOR, BOOKS
Anna Kaziunas France

LAB MANAGER
Marty Marfin

DESIGN, PHOTOGRAPHY & VIDEO

ART DIRECTOR
Juliann Brown

DESIGNER
Jim Burke

PHOTOGRAPHER
Hep Svadja

VIDEO PRODUCER
Tyler Winegarner

VIDEOGRAPHER
Nat Wilson-Heckathorn

WEBSITE

DIRECTOR OF ONLINE OPERATIONS
Clair Whitmer

SENIOR WEB DESIGNER
Josh Wright

WEB PRODUCERS
Bill Olson
David Beauchamp

SOFTWARE ENGINEER
Jay Zalowitz

SOFTWARE ENGINEER
Rich Haynie

SOFTWARE ENGINEER
Matt Abernathy

VICE PRESIDENT
Sherry Huss
sherry@makermedia.com

SALES & ADVERTISING

SENIOR SALES MANAGER
Katie D. Kunde
katie@makermedia.com

SALES MANAGERS
Cecily Benzon
cbenzon@makermedia.com
Brigitte Kunde
brigitte@makermedia.com

CLIENT SERVICES MANAGERS
Mara Lincoln
Miranda Mota

MARKETING COORDINATOR
Karlee Vincent

COMMERCE

DIRECTOR OF ECOMMERCE
Patrick McCarthy

DIRECTOR OF COMMERCE DESIGN
Riley Wilkinson

RETAIL CHANNEL DIRECTOR
Kirk Matsuo

PRODUCT INNOVATION MANAGER
Michael Castor

ASSOCIATE PRODUCER
Arianna Black

MARKETING

VICE PRESIDENT OF MARKETING
Vickie Welch
vwelch@makermedia.com

MARKETING PROGRAMS MANAGER
Suzanne Huston

MARKETING SERVICES COORDINATOR
Johanna Nuding

MARKETING RELATIONS COORDINATOR
Sarah Slagle

DIRECTOR, RETAIL MARKETING & OPERATIONS
Heather Harmon Cochran
heatherh@makermedia.com

MAKER FAIRE

PRODUCER
Louise Glasgow

PROGRAM DIRECTOR
Sabrina Merlo

MARKETING & PR
Bridgette Vanderlaan

CUSTOM PROGRAMS

DIRECTOR
Michelle Hlubinka

CUSTOMER SERVICE

CUSTOMER SERVICE REPRESENTATIVE
Kelly Thornton
cs@readerservices.makezine.com

CUSTOMER SERVICE REPRESENTATIVE
Ryan Austin

Manage your account online, including change of address:
makezine.com/account
866-289-8847 toll-free in U.S. and Canada
818-487-2037,
5 a.m.–5 p.m., PST
makezine.com

PUBLISHED BY

MAKER MEDIA, INC.
Dale Dougherty

Copyright © 2015 Maker Media, Inc. All rights reserved. Reproduction without permission is prohibited. Printed in the USA by Schumann Printers, Inc.

CONTRIBUTING EDITORS
William Gurstelle, Nick Normal, Charles Platt

CONTRIBUTING WRITERS
Lila Becker, Ayah Bdeir, Alastair Bland, Sara Breselor, Len Cullum, DC Denison, Stuart Deutsch, Steve Doll, Nick Dragotta, Sam Freeman, Austin Furey, Mark Harrison, Donald Hills, John Iovine, Bob Knetzger, Forrest M. Mims III, Laura Murray, Seth Newsome, Esterelle Payany, Matt Richardson, Andrew Salomone, Alec Short, Jason Poel Smith, Matt Stultz, Arvin Temkar, Andrew Terranova, Lucas Weakley, Eric Weinhoffer, Kenji Yoshino

Comments may be sent to:
editor@makezine.com

Visit us online:
makezine.com

CONTRIBUTING ARTISTS
Bob Knetzger, Charles Platt, Damien Scogin

ONLINE CONTRIBUTORS
Cabe Atwell, Gareth Branwyn, Tom Burtonwood, Jon Christian, Jeremy Cook, Jimmy DiResta, Vilma Farrell, FIRST robotics team The Allsparks, Agnes Niewiadomski, Luanga Nuwame, Krista Peryer, Haley Pierson-Cox, Frank Teng, Matthew Terndrup, Michael Weinberg, Tyler Worman

ENGINEERING INTERNS
Brian Melani, Nick Parks, Sam Scheiner

Follow us on Twitter:
@make @makerfaire
@craft @makershed

On Google+:
google.com/+make

On Facebook:
makemagazine

CONTRIBUTORS

What attachment would you like to mount to your drone?

Seth Newsome
San Francisco, California [Carbon Fiber Acoustic Guitar]
A boombox. By day I could listen to my jams, and by night I could command an instant street party.

Kenji Yoshino
Grinnell, Iowa [Smartphone Digital Microscope]
A bunch of laser pointers. I would use it to round up and play with all the stray cats in town.

Sara Breselor
San Francisco, California [Hands-On Health Care]
I think my personal drone would carry a heat lamp for my perpetually cold hands.

Arvin Temkar
San Francisco, California [Regal Receiver]
I'd attach a vacuum nozzle so it could get all the places a Roomba can't reach.

Lila Becker
Bellevue, Washington [Easy Mega Infinity Mirror]
I would definitely want a robotic arm attachment to be my robotic assistant in life activities such as bringing me a glass of water and my laundry.

PLEASE NOTE: Technology, the laws, and limitations imposed by manufacturers and content owners are constantly changing. Thus, some of the projects described may not work, may be inconsistent with current laws or user agreements, or may damage or adversely affect some equipment. Your safety is your own responsibility, including proper use of equipment and safety gear, and determining whether you have adequate skill and experience. Power tools, electricity, and other resources used for these projects are dangerous, unless used properly and with adequate precautions, including safety gear. Some illustrative photos do not depict safety precautions or equipment, in order to show the project steps more clearly. These projects are not intended for use by children. Use of the instructions and suggestions in Make: is at your own risk. Maker Media, Inc., disclaims all responsibility for any resulting damage, injury, or expense. It is your responsibility to make sure that your activities comply with applicable laws, including copyright.

CONTENTS

makezine.com/44

PROJECTS

Smartphone Microscope 46
Take high-magnification photos with your phone, using the lens from a cheap laser pointer.

Carbon Fiber Acoustic Guitar 50
Use free 3D design tools to mold a tough, sleek guitar body in high-tech composite.

Plasma Arc Speaker 56
A high-voltage loudspeaker makes sound waves by vibrating an electric arc.

Easy Mega Infinity Mirror 60
This quick build makes a door-sized star portal — or a vertiginous beer-pong table!

Build a Twilight Photometer 62
Experienced circuit makers can build this ultra-sensitive device to detect particles.

123 — How to Tattoo a Banana 65
Don't just play with your food — make art with it.

Pi Spy Surveillance System 66
The Raspberry Pi B+, Pi Camera Module, and MotionPi make it easy to keep an eye on things

Cloning the Fig 68
Don't steal fruit from your neighbor's fig tree — clone it!

DIY "Nutella" Spread 70
Make your own chocolate-hazelnut paste that's better than store-bought.

Howtoons 71
Make butter with reverse emulsion.

Getting Started with LittleBits 72
The new book from Maker Media will initiate you into the ecosystem of the popular magnetic components.

Remaking History 76
Re-create the process that gave us car tires, rubber bands, and the boots on your feet.

Toy Inventor's Notebook 78
Make a tiny toy Cartesian diver.

SKILL BUILDERS

Understanding Basic Woodworking Tools 80
Six simple hand tools for building almost anything.

Your Obedient Servo 84
How to build a bare-bones robot cart.

TOOLBOX

Tool Reviews 88
Recommendations for unique and useful maker tools, toys, and materials.

New Maker Tech 91
On the horizon for electronic accessories.

Books 91
Text tools for your bench or bedside table.

3D Printer Review 92
We take a look at the Printrbot Simple Maker's Kit.

Vol. 44, March 2015. *Make:* (ISSN 1556-2336) is published bimonthly by Maker Media, Inc. in the months of January, March, May, July, September, and November. Maker Media is located at 1160 Battery Street, Suite 125, San Francisco, CA 94111, 877-306-6253. SUBSCRIPTIONS: Send all subscription requests to *Make:*, P.O. Box 17046, North Hollywood, CA 91615-9588 or subscribe online at makezine.com/offer or via phone at (866) 289-8847 (U.S. and Canada); all other countries call (818) 487-2037. Subscriptions are available for $34.95 for 1 year (6 issues) in the United States; in Canada: $39.95 USD; all other countries: $49.95 USD. Periodicals Postage Paid at Sebastopol, CA, and at additional mailing offices. POSTMASTER: Send address changes to *Make:*, P.O. Box 17046, North Hollywood, CA 91615-9588. Canada Post Publications Mail Agreement Number 41129568. CANADA POSTMASTER: Send address changes to: Maker Media, PO Box 456, Niagara Falls, ON L2E 6V2

Parrot
BEBOP DRONE
SKYCONTROLLER

Own the sky with Parrot ultimate Full HD drone camera
- Lightweight and robust design built-in with safety in mind
- Full HD video stabilized on 3-axis achieved by GPU
- 14 Megapixel and Fisheye camera
- First Person View piloting
- Control the angle of the camera with the piloting application
- Extended range with Parrot Skycontroller for immersive controls

 FreeFlight 3 is available for free

From $499 - More details on **www.parrot.com**

Parrot SA - RCS PARIS 394 149 496.